Anti-Terrorist Security For Business and Industry

Loren Jackson

Bloomington, IN  Milton Keynes, UK

authorHOUSE™

AuthorHouse™
1663 Liberty Drive, Suite 200
Bloomington, IN 47403
www.authorhouse.com
Phone: 1-800-839-8640

AuthorHouse™ UK Ltd.
500 Avebury Boulevard
Central Milton Keynes, MK9 2BE
www.authorhouse.co.uk
Phone: 08001974150

First published by AuthorHouse 8/9/2006

ISBN: 1-4259-3966-X (sc)

Printed in the United States of America
Bloomington, Indiana

This book is printed on acid-free paper.

<u>Dedication</u>

This manual is dedicated to the victims of 9/11 and their surviving families. It is my feeble contribution to the War On Terrorism and a tribute to those who so tragically died. Their spent lives finally awakened our government to a war that has been waged against us for the past twenty years...which we chose to ignore. The events of 9/11 are a terrible price to pay for a "wake up call."

Loren Jackson

Acknowledgments

A manual of this nature would not have been possible without the help and input of many people.

My eternal gratitude goes to those who put in the time and effort to make this manual possible—the FBI Agents who read and gave it their stamp of approval; John Kochel, who helped format and proof the work and also designed the cover; Bob Galt who took the time to proof the manuscript and provide recommendations as to content; Evaristo Montes for his help and critique; and Walter Viney for his comments and input.

The Department of Homeland Security is also commended for the job they are doing in combating terrorism. In addition, we wish to thank the U.S. Army Information Department, instrumental in providing us with material for various sections herein.

The editors of USA Today should be acknowledged for bringing the state of private security to the forefront of the nation. The United States of America is in a very vulnerable position at this time in our nation's history and every effort must be put forth to stop terrorism—in this country and abroad. If we don't stop it before it becomes deeply ingrained into our way of life, we will be faced with the same predicament Israel has dealt with for the past fifty years. Not only is our safety in jeopardy but our civil liberties as well. It will take all of us working together to stop terrorism.

As a devout student of over thirty years in studying ways to combat terrorism, one thing has been made painfully clear to me—to stop terrorism takes a people as resolute in maintaining their way of life as the terrorists are in taking it. Weakness and fear have no place in a battle plan to win.

It is the hope of this work to open eyes and move to action in directing the business community and our industrial complex to become stronger, safer and better prepared to face the new enemy of the twenty-first century. I also hope people will come to realize that the realm of private security must be totally reconstructed and molded to fight an enemy that never sleeps and one that will never meet you on an even battlefield.

Loren Jackson

Table of Contents

Introduction

This manual was written for several reasons. It is of course a direct result of September 11, 2001. But it also is a condemnation of our private security system, a manual that will help you (1) understand the nature of terrorism to a much clearer extent, how to best (2) defend your business or industry against possible terrorist attack, and (3) introduce you to a Company (The Knights O' the Round Table) sworn to defend the United States of America against terrorist attack.

Since September 11 a new department has been created within the United States Government to exclusively fight terrorism—the Department of Homeland Security. It is a new department, concept-driven, with a new focus on terrorism and a still developing agenda. It is true that progress has been made in airport, harbor and border security. However, this deals with border incursion and air travel only and says nothing about the "sleeper cells" that have been inside this country for years and have had time to develop sophisticated lines of communication and recruit others into their terrorist ranks. Let's face it—it will take years to develop an effective anti-terrorist department where state and local governments will feel positive effects. Don't get me wrong, I'm not degrading Homeland Security by any means and am in full support of its endeavors. But we are twenty years late in fielding a force dedicated to fight terrorism at its roots and in today's climate our resources are more reactionary in nature than preventive.

But like all new agencies that develop over time, especially those fighting a new, different kind of war and in a new environment (within the borders of the United States), it will take years for it to become fully functional and effective.

The FBI has been put in the forefront of battling terrorism but is overworked and understaffed with local and state law enforcement overburdened fighting everyday crime and the FBI is not adequately trained to fight terrorism (has never been a significant part of a new agent's training agenda). So who is left to protect your business or industry, your employees, your income, and your property from a terrorist attack?

Right! The answer is **YOU** and **YOUR** security staff! At the same time, keep in mind that private sector security, from the time of its inception, has been geared and trained to fight crime not terrorism. In fact, after September 11, I did two anti-terrorist surveys, one for a bank and another for an engineering company (that housed and made repairs on naval vessels). Both surveys showed major flaws in deterring terrorist attacks, all things from inadequate lighting to improper and ineffective use of surveillance cameras, and even to the location of the security building. That said, the biggest liability was private security guarding against terrorism. Point: When the private sector security system was initially set up, it functioned to stop criminal activity—not a terrorist threat. So the primary reason for this manual is to provide insight into how to develop a security system geared to deter terrorist acts upon your business or industry.

The information provided in this manual is not a solution to terrorism or a guarantee that you will not be subject to terrorist attack. It will, however, give you some insight into terrorism and how best to create a safer business environment so terrorists may bypass **YOU** for a more vulnerable target, **THE OTHER BUSINESS** across the street.

It will also tell you how to handle a bomb threat call, how to conduct a bomb search and evacuation, what to do if a terrorist attack occurs, proper lighting of your premises, and the accurate use of surveillance cameras. An additional reason for creating this manual is to bring to light the dismal state of affairs regarding private sector security in this country.

Since September 11, I have tried to get the training, licensing and qualifications for private sector security agencies changed in the state of Virginia. I have talked to delegates, sent a brief to the Lieutenant Governor and even had a phone interview with Sixty Minutes of CBS

Television (who called me), while preparing a segment on this very subject. The fact is that private sector security agencies have become a liability since September 11 and must be changed for the sake of our economy and citizenry. I have gotten nowhere talking to politicians and have been stonewalled at every turn. Winston Churchill probably summed it up the best when he said, "The difference between a politician and a statesman is that a statesman works for the good of the people, a politician works for the next election."

The politicians in Virginia have no idea of the state of private sector security and could care less. They are more concerned with issues that will have no meaning at all if a terrorist attack occurs, such as roads, buildings, taxes and the next election. The safety of our citizens and business or industries is way down on their list of priorities. The fact is, private sector security outnumbers law enforcement five to one, yet almost anyone living and breathing and over eighteen can get a private sector security license, and these people receive almost no training at all.

Included herein is a copy of the brief I wrote, along with an article that was the front page story of USA Today, coming out eighteen months after the brief was written.

My appeal is—show your concern and contact the delegates, representatives, or congressmen that represent your state and voice concern about the abysmal condition of private sector security. This will tell you immediately if they are statesmen or politicians. To find out just how ill trained and inept your state and local security is, talk to them and, emotionally, be prepared for the downside, which is no action at all. Ask them what kind of incendiaries are they familiar with, what kind of detonators can they recognize, how would they conduct a bomb search, how would they conduct an evacuation and where would they have employees go in case of a bomb threat or actual discovery. In most cases, the answers you'll hear will be their own, not something they were taught as part of any certification or statewide policy. I have been licensed in twenty-one states and I can assure you that their training procedures are all the same. Private security is in the realm of each state's perusal—they control it, regulate it and set the guidelines. The shame of all this is, since September 11, no state has changed its laws or training procedures—yet we have private sector security guarding schools, libraries, business, state buildings, federal buildings, power companies and other entities that are subject to terrorist attack. In light of this condition, the advice I usually give companies (the only advice that makes any sense) is to hire their own security force and train them adequately. In that way they have a vested interest in the welfare of the company and employees.

A not insignificant third reason for this manual is to offer a viable alternative to nothing (currently in force within the private sector) to help you fight terrorism the best you can. A hand's on training program, conducted by a company known as "The Knights O' the Round Table," first conceived as a Strike Security Force and Executive Protection Specialists' group. After September 11, they saw the threat of terrorism on our shores and the need for anti-terrorist protection, consulting, training and advisement to business and industry in the United States. To that end they have assembled a group who specialize in Anti-terrorist Security. They function in several capacities, from doing an anti-terrorist survey onsite—coming in and evaluating your security measures and advising what changes need to be made, from those things obvious to lighting, surveillance cameras, present security and to set up an evacuation route and procedure plan. They also provide a turnkey operation from training the business receptionist on how to handle bomb threat calls, to what your executives should be doing in the event of an emergency. Instruction is provided regarding computer security, training in-house security over and above state requirements, including baton and OC (pepper spray) and firearm certification (if your company, business or corporation uses unarmed security, you are automatically high-risk) according to state regulations. Proper K-9 support and training is a must, and if you want to form your own security force they will screen, hire, and train them in anti-terrorist security measures.

The Knight's O' the Round Table are dedicated to preserve, protect and see to it that nothing threatens our life, liberty and pursuit of happiness. At the end of this manual, I will include a biography of the principals and a list of e-mail addresses, FAX and phone numbers you may contact. They can provide an invaluable service that our government can't provide at the present time.

The importance of protecting our business and industry against terrorist attacks cannot be overemphasized. Disrupting business and industry are one of the goals of terrorists. Have you thought about what a terrorist attack would mean to you and your business? Do you think it can't or won't happen to you? If you're unprepared and it does happen, first up is loss of property and cash flow, and this is of minor importance to what follows—collateral damage (injury and loss of life to innocents) and lawsuits. Any rookie lawyer in America could prove you did not have adequate security to prevent a terrorist attack. The loss of property pales in light of the residual damage through lawsuits and insurance implications. I frankly believe that insurance companies, at some point will factor in a terrorist clause in their policies if we are hit often and with impact. To date, this has not happened. But when is the next September 11? I also believe insurance companies will factor in a discount for businesses which have taken adequate anti-terrorist security measures. In this scenario, you have done everything in your power to prevent a terrorist attack and any court in the land would hold you blameless in a lawsuit situation.

Three things you can do to protect yourself and your business:

1. Digest and FOLLOW the guidelines in this manual.

2. Have an Anti-terrorist Survey done and FOLLOW the recommendations.

3. INSTALL a complete turnkey operation.

Anything you do will be at a cost, but you must weigh the consequences of not acting. One way or the other, you will benefit from the information provided in this manual. It is our hope you will use it to your advantage and to the safety of our people and nation.

Loren Jackson

1

State of the Union

WHAT COULD AMERICANS HAVE DONE to prevent 9/11? A ccusations have been leveled, including a lot of finger pointing directed at various national security agencies, even to The White House. According to a congressional report released in July of 2003, the CIA and FBI share the blame for failing to "disrupt" the 9/11 attacks by keeping would-be terrorists out of the country or trying to unravel their sinister plot at the grassroots level. This nearly 900-page joint House-Senate project stands as the most extensive report yet compiled about what the government knew and didn't know of bin Laden and his plan to attack the United States.

Whatever the reason and whoever was responsible is irrelevant in today's climate. None of these studies can rebuild the twin towers of the World Trade Center or bring back those who died so tragically. Our focus now should be to make sure it doesn't happen again. Unfortunately, Americans, as a people, have short memories. The Emergency Response and Research Institute's (ERRI) national security analyst, Clark Staten, voiced concern over the American public developing a "complacent attitude" towards terrorism directed against the United States as if, by ignoring the matter, it might go away. "We are concerned that as each day passes without a major incident within the United States memories of September 11 will fade from everyone's consciousness..." Staten commented. He went on to say," Our ongoing analysis of the available evidence would suggest that this is not a good time for the American public to let down its guard… In our considered opinion, there is still danger on the horizon."

Staten urges continued vigilance and awareness on the part of the American public and the exploration of all U. S. efforts to best prepare for the possibility of a future cataclysmic event at the hands of terrorists. I, for one, agree with his statements and feel we have not yet felt the full impact of terrorism or the destruction and human suffering it can cause. You must remember, we are not dealing with soldiers, rather with homicidal maniacs who will destroy themselves in order to bring death and destruction to others.

The government has already acknowledged the existence of terrorist cells within the United States that are just waiting for the word to become operational. Deputy national security advisor Stephen Hadley has said, "We don't know how much they have in the can. What we worry about is that there are operations already trained, populated, planned and funded, and they are simply waiting on an opportunity." U. S. officials and terrorist experts believe that the elimination of al-Qaeda wouldn't stop the growth or agenda of militant Islam extremists whose only goal is to bring death and destruction to Western countries. A good illustration of this is the case of Richard Reid (a British national and recent convert to militant Islam) who boarded a jetliner from Paris to Miami with his sneakers filled with explosives.

The Reid incident, a man whose only crimes up to that time had been as a petty criminal, shows that independent terrorists may be plotting attacks with little or no support from al-Qaeda. Eric Davis, a terrorism expert at Rutgers University, puts forth another problem in the war on terrorism. His theory is that al-Qaeda is a cell organization decentralized worldwide. This leads to a lot of confusion and uncertainty about their actual size and scope.

Terrorists can say they are part of al-Qaeda when they are not, and this may give the illusion they are stronger than they actually are. The whole point being that people who subscribe to the same views as al-Qaeda may act on those views. It is not simply al-Qaeda we are dealing with but terrorism as a whole and worldwide.

Should we assassinate or kill bin Laden, it may only enhance the terrorist's mind-set. Bin Laden has set himself up to die a martyr and the Islamic terrorist will thrive on that concept no doubt. Moreover, the death of bin Laden will escalate the threat of terrorism in the United States in the form of reprisals. We have not had any major incidents since September 11, which is good news, and something to ponder—every day we have no incident, we have a chance to become stronger and more effective in fighting terrorism. This does not mean that terrorists have been inactive however. At least a half-dozen plots, attributed to al-Qaeda, have been thwarted, including a plot to blow up the U. S. Embassy in Paris and plots to attack U. S. interests in Singapore and Yemen. These incidents are a matter of record.

Even though these incidents were overseas we should not be complacent. The United States is what they want to destroy and what better way to attempt that than inside our country.

On April 17, 2002, when ERRI's Clark Staten reviewed the tape released by the al-Jazeera Television Network featuring Usama bin Laden and Ayman al-Zawahri he stated, "In the tape, bin Laden reportedly trumpeted the negative effect that the attacks on the World Trade Centers had on the U. S. stock markets." He added, "This statement by bin Laden is consistent with the theories and concerns of asymmetric warfare we have put fourth previously." He was referring to a 1998 report on "Asymmetric Warfare, the Evolution and Devolution of Terrorism." In this report Staten proved to be prophetic.

In the report Staten said, "More enlightened terrorists have discovered, or will discover soon, that the path to the fear and chaos they crave most may be more easily achieved by a wide-scale attack on infrastructure/ economic targets." The time has come then for U. S. business and industry to realize that "bad guys" also like "soft targets" and that as our military and government facilities become more "hardened" terrorists are increasingly likely to undertake attacks on commercial concerns.

Both the public and private sector must work together to build a defense that encompasses all of the vital infrastructure that makes America the premier superpower of the world." If we had taken heed of this *a priori,* we may not have had a September 11.

We are in a war (make no mistake about it), a different kind of war (no uniformed enemy, no innocents, no foreign country, and no quarter). We must work together. Most of all we must be diligent in providing the best security possible by using the best in manpower and technology to deter terrorism. This is absolutely necessary to keep our business and industry fruitful and out of harm's way.

This new war will be our antagonist in the 21st Century. More alarming, I do not see a war involving countries or nations, but an internal war, involving a misrepresentation of the Islamic Religion and Western Civilization at-large. This is not a scare tactic. But if we allow them to scare us, they have won a victory. We must realize this is a different war, and we must understand the nature of the enemy and how he fights. He is fanatical, motivated and trained. He is also on our soil and very, very dangerous.

2

Terrorism

HISTORICALLY, TERRORISTS OPERATE IN THE country where they plan to carry out their acts of violence. The menace strikes when least expected and the threat ever-present. Terrorism has been around for centuries and, up until the turn of the last century, terrorists fought with a code that recognized innocents like women and children. The real purpose of terrorism today is to create fear among the survivors. For this reason they operate in small groups called "cells." Previously, terrorists worked from a hierarchy system with one leader dispensing orders down the ranks on plots, agendas and intentions. Today, they seem to have gone to a decentralized mode, making it harder to infiltrate, uncover plots and also harder to extract information from captured terrorists because they operate only on a need-to-know basis. Cells are generally responsible for only one phase of an overall plot and have no knowledge of the rest of the intended act or who is involved. In this manner, a terrorist involved in a plot that is captured can be replaced with little or no adjustment to the intended plot (larger picture).

On September 18, 2002, U. S. officials confirmed to the Washington Times that information obtained from self-professed 9/11 organizer Ramzi Binalshibh indicates al-Qaeda has decentralized its leadership structure, making it more dangerous. The report also said that terrorist cells now have more autonomy to conduct attacks around the world without support or authority from the hierarchy. One U. S. intelligence official confided to Times reporter Bill Getz: "The decentralization means the danger from this group is growing." Magnus Ranstorp, deputy director of the Center for the Study of Terrorism and Political Violence at the University of St. Andrews in Scotland concurs. He maintains that terrorism is directed from the bottom up as much as the top down. The typical pattern before September 11 was of local al-Qaeda cells initiating reconnaissance of potential targets, planning and then going back to al-Qaeda leadership for approval and possible funding. The foot soldiers today are "self-initiating and self-sustaining."

Decentralization of the terrorist hierarchy and letting local cell groups operate independently presents a multitude of problems, especially in the U. S. where, for years, we have allowed Moslems sanctuary without question (this issue will not be solved overnight). We have a large number of Moslems from the Hamas and Hezbollah sects, two of the largest militant groups. It will take time and patience to ferret out these cells and put an end to their ability to effect violence.

Even though they have decentralized and given more autonomy to the local cells, their recruiting, training, goals, agenda and methods of operation have essentially remained the same. Recruiting primarily comes from dissident students, ex-military and religious fanatics who openly speak out against a "decadent west." Recruiting in the past has been mostly from the middle-eastern countries but there is clear evidence that, for years, recruiting has also taken place in the U. S. and Canada, as hard as that is to admit. Those who are chosen are carefully screened and watched before they are actually approached for enlistment. The recruitment process is such that it is almost familial in terms of recruitment. Defection or betrayal is rewarded by death, not only for the betrayer but his entire family as well. Because of their screening process, methods, and end results, it is extremely hard to infiltrate their ranks as we did with Russia during the Cold War. Their training takes place in other Moslem countries. The program usually consists of small arms and automatic weapons training, explosives, their construction, detonators and detonating devises. Intelligence gathering, photography and a literal brainwashing in hatred for the west and indoctrination to the honor of dying in a Jihad (Holy War) is the central thrust in the indoctrination process, thus the recruits become, to their way of thinking, martyrs.

The ex-military are the elite and usually instructors at these camps. The trainees come from all walks of life and are thoroughly trained. These are the "expendables" that are willing to die perpetrating acts of violence, through bombings, assassinations, suicide and anything to bring death and destruction to the "western infidels." After training, they are dispersed to different locations, given a "point of contact" and adopted into a cell.

Once they reach their appointed cell they are further indoctrinated into the importance of the "cause" and the blessings they will receive upon becoming martyrs, dying for Allah and bringing destruction upon the enemy. Thus, they become fanatical, suicidal and extremely dangerous. Up until they decentralized, a terrorist plot could be exposed by the capture of a major player or cell. Now, one cell only handles a particular phase of the operation and the trainees are not given a timetable or location until it is time to strike, with the pieces only then fitting into place. For example, using September 11 as an illustration, it has been guesstimated that all principals involved in the hijackings were trained in different locations, didn't know one another and never given a location or timetable.

The terrorists in charge of taking control of the hijacked 9/11 passenger aircraft were trained in that aspect and that aspect only, while the pilot training took place elsewhere and they were given only information as to their part in the event. When it came time to act, the plot came together just as you would put a puzzle together—all parts in place.

Using this type of tactic it is no small wander that our security did not uncover the full scope of the plan but had only fragmentary pieces that would not fit into a whole to project any kind of primary aim. September 11 was the greatest single coup ever perpetrated on the United States of America, even including Pearl Harbor. This is so primarily because it achieved every goal that terrorists wanted to achieve in their acts of violence and then some. Point: No one could have predicted beforehand that the twin towers of the World Trade Center would collapse upon impact from aircraft.

In general, the total impact terrorists wish to achieve is:

1. A significant loss of lives and property.

2. An undermining of local and national security and government.

3. To show the world and other terrorist groups we are vulnerable.

4. To disrupt our economy.

As you can see, al-Qaeda achieved all these goals in one strike. In this era of terrorism, the perpetrator has much more at his disposal than terrorists of the past. The availability of weapons and explosives on the "Black Market" (especially after the dissection of Russia), the sophisticated timing and detonating devises of today, global communication techniques (e.g., computers, internet and e-mail) add an infinite capacity to the terrorist threat. The world stage has lent to a new form of terrorism I choose to call "Shock Terrorism." The practice of kidnappings, assassinations, hijackings, and small-scale bombings will continue as in the past but only to our rivet attention to those specifics while the perps plot acts of greater magnitude elsewhere, such as September 11.

In July of 2002, an assessment at ERRI suggested that lower level al-Qaeda associates were in the midst of planning a series of small or medium strikes on "easier" targets...to include car bombings and assassinations. Independent cells will undertake these acts with motivation by, and possible funding, but with little direction from the top leadership of al-Qaeda.

ERRI's assessment is that we will see this trend continue for some time with a greater than even chance of an attack of much greater magnitude, possibly involving Weapons of Mass Destruction (WMD) in the future. One thing should be perfectly clear—there is no plant, business, industry or building that is not accessible by, and a potential target of, terrorist violence.

Terrorists have many advantages over our security and law enforcement agencies. By not recognizing innocents, any and all targets are fair game and they determine their targets and decide when, where, and how to attack—at their leisure. With all of these choices it gives a terrorist good probability of success with minimal risk. Even if there is a maximum risk, these people will not hesitate to die carrying out their acts.

Terrorists employ a variety of methods for assessing viable targets. The following seven steps are usually among the short list:

TARGET CHECK OFF LIST

1. **Target Selection:** Accessibility, level of security and the projected total effect of an attack usually determine the game plan.

2. **Surveillance**: The first surveillance will be informal and casual. Watching for weaknesses, observing security (are they armed, how often they make their rounds, do they perform their duties proficiently or halfheartedly, what time the guard changes, etc.) and weighing the probability of target acquisition.

3. **Planning**: Determine what weapons are needed, how much manpower is needed, the best time to strike, the access route and the escape route.

4. **Final Target Designation**: This will be based on reports from the first three steps. The easiest target with the promise of the most damage will be selected.

5. **Details**: This will include the selection of individuals needed for each phase of the job and his exact duties. All logistic

measures are covered here as well as housing, transportation (etc.).

6. **Final Surveillance**: This is a much more professional and detailed surveillance than in step one. This includes photographs of the target (aerial when possible), maps of the area, and any changes in security or procedures, also, if available, schematics or blueprints of the target. This enables them to make further adjustments in steps three and five. At this point a timetable is also put into effect.

7. **The Actual Attack**: Somewhere within these seven steps your security must be able to spot and take steps to curtail a terrorist act. By security, I not only mean your human security, but your cameras, lighting, and accessibility security (covered in the next chapter).

TERRORISM AND ISLAM

Following the tragic attack of September 11, CNN conducted a survey of Americans about how much they knew about the Moslem Religion. Sixty-four per cent of the Americans polled admitted they knew little or nothing about the religion of Islam. As a result of this ignorance many people associated Islam with terrorism and reports of hate crimes, directed toward Moslems, were being reported. The true Moslem religion is in no way connected to acts of terrorism. On the contrary, the precepts of the true Islamic faith are stricter more than those of the Christian faith. True Muslims don't drink, smoke, or tolerate lewdness in any way. They are considerate of others, charitable, and well meaning people. Terrorists use the Koran and Islam to hide behind and justify their violent acts. Karen Armstrong, an authority on Islam and author of the best seller, "Islam: A Short History," was asked in an interview by Newsweek Magazine, "We've all heard that suicide bombers believe they will go straight to heaven and enjoy a paradise of milk and honey, with 72 virgins for every martyr. Is there any religious basis for this?" Her answer was quite forthright, saying it was completely illegitimate. She also added, "The Koran and Islamic Law forbid suicide in the strongest terms. Also, according to the Koran, Muslims must not wage a war against a country where Muslims are allowed to worship freely. Islamic Law also forbids the killing of women and children." But was Saddam Hussein truly a Muslim, if we exact this definition?

So, it is a false version of Islam to imagine these suicide bombers are buying a first class ticket to heaven. That said, then how can a completely innocent, gentle religion become a background for such violent acts against humanity, including women and children? First, you must remember that in the Middle East most people are poor and uneducated. Most of them lack the most fundamental understanding of Islam—even to the local level. For the most part they rely on Sheikhs, self-proclaimed scholars and even terrorist leaders for teaching the precepts of Islam. These "teachers" mold and bend Islamic law to their own convenience and agenda. In most all cases, the version they offer is totally different from true Islamic Law and the teachings of the Koran, focusing instead on the roots of political and social confrontation. To this add the "fatwahs" (religious proclamations) of the self-proclaimed teachers to the existing, twisted, version of Islam and you have the makings of a terrorist situation. This is how a terrorist like bin-Laden can have great influence and control over those Muslims whose understanding of Islam has been taught by slanted, self-serving teachers. In this fashion, bin-Laden has the ability to mold, shape and philosophically direct them to do his bidding.

A terrorist may proclaim Islam as his religion but a true Moslem is not a terrorist. However, as it is totally wrong to look at Islam as terrorist oriented it is also hard to distinguish a true Muslim from a terrorist Muslim because the terrorist hides behind Islam to conceal his true intentions.

THE MOTIVATION OF TERRORISM

According to the U. S. Army Field Manual 100-20, Stability and Support Operations, Chapter 8, Combating Terrorism, the act of terrorism is classified into three categories:

1. **Rational:** The rational terrorist will think. He will weigh his chances for success in much the same manner one would look at a profit and loss analysis. The rational terrorist will look at his intended act to see if the action will accomplish its purpose and he will also consider the downside, in other words, will the act have the desired repercussions as opposed to the cost.

2. **Psychological:** The terrorist who is psychologically motivated is possibly the most dangerous, especially when combined with religious fervor. His only reason for existence is terrorist action. He will impart his mentality to anyone who will listen, thus recruiting those of the same mind-set, enabling him to dehumanize his victims and kill without guilt or remorse. The psychologically motivated will lure disciples who are often young, social misfits, of little or no education. These individuals typically find their group acceptance in community gangs and their relief and gratification through the use of drugs and violence, thus becoming easy targets for terrorists who offer much needed acceptance. The hierarchy of this kind of terrorist may be single-minded, but not stupid. They will use these misguided outcasts as pawns to further their cause.

3. **Cultural:** The culturally motivated terrorists can evolve from fear of extermination, thus finding a means of survival through terrorism. The Viet Cong and the Afghans are prime examples of this in the twentieth century. As we have experienced with al-Qaeda, combine rational terror with psychological terror backed by extreme cultural and religious fervor and disaster is eminent. The Moslem extremists have exhibited each and every one of these attributes.

As you can see, terrorism is a different type of violence. The threat is ever present and **THERE ARE NO INNOCENTS.** The danger of terrorism is clearly that—**THERE ARE NO INNOCENTS.** The Department of Defense (DoD) has defined terrorism as "an instrument that uses calculated use of violence or the threat of violence to incite fear, intended to incite fear, to coerce or intimidate governments or societies in pursuit of goals that are usually political, religious, or ideological." Their goals are usually specific, well planned with a well thought out effect of the act.

WHY US?

According to expert opinion, Islamic extremists have developed their philosophy into a cultural war. Seemingly, the presence of western culture imported to the Islamic Nations was unacceptable by regimes unable to merge with the political and cultural perceptions of the "decadent" West. The outcome of these differences was a development of Jihad (Holy War). This offered a clash between the civilizations of cataclysmic proportions. One of the greatest accomplishments of the terrorist groups was their ability to present this ideology to a generally uneducated Muslim populace as an inexorable clash between Islam and the West. The next stage was to convince the Muslim population that war is constant between the Islam and the West. The feeling of being in a war of self-defense appealed to a large number of Moslem followers. Now Jihad not only means a military conflict but a cultural one, as we are a perceived threat to them. The result of this was a new generation of Islamic terrorist—defenders of Islam in this global war.

In the late 1980s al-Qaeda was formed by Osama bin Ladin. His objective was to develop a pan-Islamic Caliphate all over the world by uniting all Islamic extremist groups to overthrow any governments it deems non-Muslim and expelling Westerners from Muslim countries.

In a statement issued in February 1988, he said it was the duty of all Muslims to kill U. S. citizens. An inkling of what was forthcoming should have been gleamed from the book, "America From The Inside" (Amerika min al-Dakhil), written in 1952 by Sayyed Qutb, in which he predicted the oncoming threat of global terrorism, in the name of Jihad. Now bin Laden is the undisputed leader of the militant Moslem faction worldwide.

If we kill bin Laden, we make him a martyr (it won't destroy al-Qaeda), if we take him prisoner, we must prepare for war on a wide territorial table which includes the U. S. They would stop at nothing to free their "savior." If we kill him, it will leave a legacy and someone else will take the reins. We are at war and must prepare for it.

3

Fighting Back—Anti-terrorist Security

Anti-terrorist security is a relatively new concept that, before September 11, had been relegated to military bases, government buildings and nuclear sites. Now Anti-terrorist Security must apply to every business, industry, building and person in America. You must be constantly on the alert 24/7/365 for anything out of the ordinary and duly report to proper authorities. That's just how it is and will be tomorrow.

September 11 did more than take lives and destroy two prominent landmarks. It brought to our attention the fact that we are vulnerable as a nation. Our security was lax, our intelligence insufficient, our borders susceptible to entry. As a nation, we were caught unaware. Our entire infrastructure was shown to be in grave danger. Our immediate reaction was a counter-terrorist response, war in Afghanistan—in other words, take it to the enemy. The difference between counter-terrorism and anti-terrorism is, **counter-terrorism** is a reaction (usually violent in nature) to a terrorist act after it occurs, while **anti-terrorism** is a concerted effort to prevent any terrorist act from occurring. While counter-terrorism exacts retribution, it is certainly not the most feasible route to go. It's total concept in execution is based on the "an eye for an eye" precept. Counter-terrorism costs lives while anti-terrorism saves lives. It certainly does not take a genius to figure out which path we should follow.

The problem lies in the fact that our internal security is trained for criminal activity, not terrorist activity. Our law enforcement agencies, from the state police down to the local level and private security have had little, if any, training in anti-terrorist security.

Granted, there are some similarities between criminals and terrorists, but there are some major differences as well. One difference is the scope of act—a criminal's actions are usually done for personal gain and limited in scope (e.g., burglary, robbery, murder). A terrorist's goal is the death and destruction of lives and property in a massive agenda. Another big difference is that a criminal wants to commit his crime and escape, meaning he must get to the target of his crime and then get out, hopefully without detection. A terrorist just wants to acquire the target, getting away is of no consequence to him. If a terrorist penetrates the perimeter of his target, be it a building, plant, restaurant, or airplane, odds are, he has already achieved his objective and someone will die. So, essentially, the primary goal of anti-terrorist security is to keep the terrorist from penetrating the target perimeter or to uncover his intentions before he has a chance to act.

Targets fall into two categories, the "hard target" and the "soft target." The hard target is one that is at a high security maintenance level. In other words, their guards are usually armed and professional, lighting is more than adequate at night and cameras cover a major portion of the property or building. The perimeter is fenced completely around, an adequate alarm system is in place, entrance and egress is carefully monitored, building and grounds patrolled. Periodically identification checked before allowing access to property. On the other hand a "soft target" is one that has easy accessibility, poor security personnel, very few, if any cameras and shows that target acquisition can be accomplished at minimum risk, inflicting maximum damage. Of the two, the "soft target" will always take precedence over the "hard target." The terrorist will then assess the value of hitting a particular target, weighing the damage he can do and how much media exposure it would create.

The immediate priority of any business or industry, then, is to become classified a "hard target" so if they are being considered a target of terrorist activity in all probability terrorist cells will bypass that business activity in favor of a softer target where a greater probability exists for inflicting damage and getting away with it. Don't get me wrong, being classified a "hard target" doesn't guarantee you won't become a target. If terrorists think you are valuable enough, they will make you a target regardless and will go to great lengths to achieve a successful hit—even though it means certain death to them. However, being a "hard target" has the advantage of discovering a plot before it ever happens because of the in force high-level of security.

Through the use of surveillance cameras, properly monitored and security personnel (if properly trained to know what to look for), you stand a good chance of spotting a surveillance activity on your business or building and can notify the proper authorities. All of the advantages lies with the "hard target," as overwhelming security is displayed—in surveillance cameras, alarms, lighting, impenetrable access, professionally trained security guards, ID checkpoints and vehicle searches making access difficult.

Defending against terrorist attacks is a 24/7 proposition and a periodic check of all new anti-terrorist techniques and high-tech equipment available should be investigated periodically.

Anti-terrorist security is costly and time consuming, but considering the consequences and possible losses in lives, property and down time for your business, it is well worth the effort. In the next chapter we will look at the importance of proper lighting schemes for your business and the invaluable use of surveillance cameras.

4

Lights, Cameras, Action

THE WHITE HOUSE PRESS CORPS and the President of the United States, unfortunately, immediately after 9/11, began referring to terrorists as "Shadow Warriors." This description and ensuing implications lends the wrong directive. Though it's true, terrorists dwell in the shadows and thrive on darkness and anonymity to carry out their surveillance and deeds, calling them warriors is slighting every honorable fighting man in the world. A warrior has a code he lives by, a conscious, and a set of rules he applies in everyday life and in war and abides by such precepts at all times. Terrorists have no code, no compassion, no mercy. They kill without discrimination, remorse, and they target all people—men, women, children, the elderly, the incapacitated, as "there are no innocents." These are not the actions of warriors but of fanatics, who value nothing social, political, or religious and do nothing for the betterment of mankind. However, they are shadowy and use darkness to hide their surveillance in a given project.

The first order of business in anti-terrorist security is to make sure your premises are well lit. This includes your building, fence perimeter, parking lot and any other dark area around your building and property. Remember what was said in outlining the steps a terrorist uses in planning an attack. The first is an informal surveillance to determine accessibility and security of a target. This surveillance is done both day and night, but surveillance during the day is usually done at a distance to avoid detection. At night, if a target is not well lit, the surveillance becomes up-close and personal. Plus, a dimly lit target makes it more feasible to launch a nighttime attack under the cover of darkness. A tremendous amount of information can be acquired and acted upon if your property does not have adequate lighting and allows the terrorist surveillance team to get close and observe your entire operation.

They can monitor in-coming and out-going traffic and how well the business checks IDs. They can spot and mark any camera activity within the property, can spot all entrances and exits to the property, and they can get an accurate, personal glimpse of your security personnel. How often they make rounds, how thorough are they when they make their rounds—do they check all doors, do they check trash containers, do they check dark corners and do they investigate unusual objects? How perceptive they are, their demeanor (do they do a professional job or just "make the rounds") and down to what time the shift changes. Granted, a lot of this information can be gathered from a distance and during the day but not nearly as thorough as at night. Let me give you an example of a prime terrorist target that I happened across during one of my surveys. It was an engineering company, which does repairs on U. S. Naval vessels. Two navy vessels docked for repairs. The parking lot, adjacent to the administration building, was totally without lights (absolutely none) and at night you could not see from the security building to the parking lot. Even though the parking lot was used for naval personnel (those stationed on the ships) anyone had access to the parking area. To the point as well, the guard building was located some twenty yards inside the gate.

All personnel walked into the perimeter of the property twenty yards before having to show ID. If that wasn't bad enough, ALL vehicles were allowed into the premises, without question, through the gate by a remote from the security building. That being so, of what purpose was the electronic gate? Visitors drove up to the security building where they would encounter a stop sign and were supposed to stop, state their business and go through a vehicular search. Another faux pas—while doing the facility on-site survey, the Post Commander one day asked,

"How do I recognize a bomb or explosive?" All of this can lead to an attack and is a disaster waiting to happen.

A terrorist not only has free access to observation from a totally dark parking lot; it would not take a rocket scientist to see that all vehicles were allowed access to the property, without hesitation and without question. The only thing stopping a truckload of explosives from driving through the gate and reaching the ships directly behind the gate about forty yards was a stop sign. Do you really think a terrorist bent on blowing up a ship is going to observe a stop sign? Fortunately, this location was a remote place but if the terrorists ever get a line on it, an attack will be imminent. Their security is so lax it rates lower than a "soft target." Yet our naval ships are being housed at this location. The U. S. S. Cole, you'll remember, was hit by a vessel in the water, bent on attack when it couldn't miss. A land-based assault on an installation with no security was be absolutely catastrophic because of the explosives able to be delivered. Some other problems I have encountered, and these are frequent ones, are burnt out light bulbs in strategic lighting areas, shorts in the wiring that cause flickering or failure and a general lack of adequate lighting in some areas.

Point: Always make sure your lighting is well maintained, bulbs are replaced and any problems are taken care of immediately. In essence, make sure your grounds (buildings, parking lots and all accessible areas) are well lit. Such precautions may not stop a terrorist attack if he wants you bad enough, but it will certainly make him think twice before targeting your enterprise and may cause them to move on to a "softer target."

CAMERAS

In the past twenty years, developments in the field of video and surveillance cameras have been phenomenal. Cameras are instrumental in deterring crime and have been crucial in the capturing of criminals and their prosecution. Well-placed surveillance cameras (not easily accessible and damage-protected) are as essential as good lighting in anti-terrorist security. Cameras should be placed where the entrance and egress of vehicles take place, rotating cameras should provide a complete view of the premises, and inside cameras should monitor hallways, stairwells and entrances and exit points. An important thing to remember about your surveillance cameras is to **make them visible**. Hidden cameras are good for recording criminals in the act of committing a crime and then identifying the perps and prosecuting them. By making your cameras visible to terrorists, you have identified yourself as a "hard target." Terrorists, unlike criminals, generally are not interested in disabling your surveillance cameras. Since they strike quickly and violently and, in most cases, the explosion will destroy the tapes of the cameras, they could care less about disabling them. But just the fact they can see them functioning makes surveillance and a possible incursion into the facility very risky. Again, this may be the instrument that will cause them to seek a "softer" target. Incidentally, the cameras surveying the property should also, if at all possible, be focused to encompass any passing traffic or parked cars across the street (in other words, other, ancillary locations). In reviewing the tapes special attention should be given to both of these areas—looking for cars making repeat passes, strange vehicles parked across the street, constant loitering on the sidewalk, etc. Always remember we are dealing with a different animal here—one who must be stopped before he strikes.

This takes a lot of patience and hours of reviewing surveillance tapes looking for anything out of the ordinary. They are not going to raise their hand and say, "I'm a terrorist." Nor will they be dressed in Middle Eastern garb. It may be a glance, movement or just a gut feeling that will tell you, this is a terrorist. At this point, proper authorities should be notified for their assessment of your observed concerns.

Good lighting and proper camera use are two excellent deterrents to terrorist activities. They have pretty much marked you as a "hard" target and most times a terrorist will move on to a "softer" target. Again, being a "hard" target does not make you immune from attack but terrorists will not attack a "hard" target unless the political, economic, or collateral damage value is very great. For instance, LAX is well lit. To say it's off limits to a potential terrorist attack is ludicrous.

ACTION

The proper action (application of security in force) is to have your business install a total package that involves integration of lights, cameras and security personnel with one goal in mind—proper monitoring.

In order to utilize both of these deterrents the end result lies in human hands. All of the lights and cameras in the world are of no value without the human factor working in tandem with effective equipment. Security personnel must be aware that lights are there to uncover any unwarranted vehicles, people, unusual objects or unusual activity. Lights are there to give them a point of view, not to make sure they can see where they are going. Security personnel must be constantly on vigil, aware, rested and extremely alert. Anything unusual should be logged and reported at shift change.

Many business enterprises have used cameras for years, but in a manner that does not work in the case of a terrorist threat. In the past, most surveillance cameras were used to record, but never reviewed unless an incident occurred, such as a break-in, robbery, theft or accidents. The whole point is, the film was not reviewed until after the fact.

In dealing with terrorism, these films must be pulled, viewed and analyzed on a daily basis. Even in cases where a monitor shows everything going on and someone is watching the monitor. You can't expect him to analyze each detail, as the camera and scenes are constantly changing. He can watch as the camera rolls, but he can't analyze any given situation. In some cases, he may have several screens to monitor and this makes it almost impossible for him to pick up anything significant that's "on the move."

It is a boring, tedious, time consuming job but something that needs to be done constantly. You can take a chance and let the cameras roll, hoping beyond hope that whoever is watching, doesn't take his eyes from the screen and can spot something suspicious. For high-risk targets, however, this could be a fatal decision. Another thing to keep in mind, insurance reimbursements never cover 100 percent of anything, and in a big bucks business a terrorist attack even after settling insurance claims can result in millions of dollars lost, irretrievably lost to the owner forever.

5

Additional Anti-terrorist Security Measures

AT THE PRESENT TIME A, lot of speculation exists on whether terrorists have access to Weapons of Mass Destruction (WMD) or, at least, the knowledge and capabilities of launching a chemical warfare attack. Until further evidence is uncovered on this issue, we must deal with what we know as facts. One fact is certain, explosives have been the trademarks of terrorists throughout the twentieth century, and this should be our immediate concern. In the past, bombs have been crude but effective in inflicting death and destruction. Today's technology has made explosives more sophisticated and deadly.

Bombs come in all shapes and sizes and can be made up of many types of explosive material (dynamite, C4, Semtex, Nitro, and even common household products; the Oklahoma Federal Building bomb used common fertilizer as a base element). Detonators too, have become varied and sophisticated. They can be a timing devise, mercurial, remote, computer, phone, and pager generated. Delivery methods vary with the personalities of the perps, from a truckload of explosives driven by a suicide-bound terrorist, to car bombs, mail bombs, well placed timing bombs, to suicide bombers, lining themselves with explosives and walking into a building or crowd and detonating himself.

Explosives are definitely our biggest threat and one that we must be prepared for and geared to prevent. We must also be prepared to act in case an incendiary devise does hit us causing damage and injury to be minimized and making shock recovery quick and efficient. This is why your lights, cameras and security personnel must be capable, diligent and competent and, above all, **trained**.

Stopping an explosive devise from getting inside your premises is a job for your security measures not the next-door neighbor's. Spotting a suicide bomber or looking out for car bombs require different techniques but still rely on the awareness and competence of **your** security people.

Spotting a suicide bomber:

1. Look for a person wearing unseasonably warm or cold attire.

2. Look for protruding bulges in clothing.

3. Look for mumbling or fidgeting.

4. Look for someone consciously avoiding police.

5. Look for someone trying to fit or hide in a crowd.

Spotting a car bomb:

1. Look for mismatched license plates.

2. Look for vehicles parked in unusual locations.

3. Look for the car weighted down in the back because of the weight of explosives.

If security personnel or anyone at your business sees something that may indicate a possible suicide or car bomb, stay calm, notify police and keep the suspect or car in sight and at a safe distance.

THE MAIL ROOM

As mentioned previously, for years, using the mail has been a method of delivery for terrorists in the form of packages and "letter bombs." Several things must be done to monitor the mailroom that will help uncover an explosive. Again, it is not foolproof, but these methods are reliable and if followed you stand a good chance of uncovering an explosive package or letter.

If the contaminated package or letter is followed by a threatening phone call follow the bomb threat data sheet in the next chapter.

Each letter or package should be checked using the following criteria:

1. See if the mail comes from a stranger or unknown place.

2. Notice a lack of return address.

3. Note an excessive amount of postage.

4. Is the size abnormal, excessive or unusual?

5. Is spelling correct?

6. Do the return address and postmark differ in location?

7. Does the handwriting appear foreign in style?

8. Does the package smell peculiar (many explosives used by terrorist smell like shoe polish or almonds)?

9. Is the package unusually heavy or light?

10. Is the package uneven in balance or is it lopsided?

11. Is there springiness to the top, bottom, or sides?

12. If the package is cardboard, is it sweating on the outside (cardboard does not sweat)?

If the answer to any of the above questions is yes, immediately report any and all seen and unseen (heard or inferred) hiccups to the security officer.

Other things to consider inside the workplace:

1. Do not tarry for any length of time near a window.

2. Place desks out of sight of windows and passers-by.

3. Avoid trips to the office when no one is there.

4. Vary arrival and departure times.

5. Use different routes and entrances.

6. Know the location of all fire extinguishers.

7. Have an exit route planned in case of an emergency.

8. DON'T BE PREDICTABLE!

6

Bomb Threat Calls

BOMB THREAT CALLS ARE NOT necessarily a trademark of the Middle Eastern terrorist but have been known to happen (particularly in Europe). However, such calls have been a characteristic of many other terrorist groups that are not Moslem (we should not forget that there are other terrorist groups that would like to see America brought to its knees) who joy in making bomb threats over the phone.

The following information should be readily available to anyone who is available to answer the phones:

On receiving a bomb threat call, you should:

1. Remain calm and record the time of the call.

2. Listen very carefully and, if possible write or record the conversation verbatim.

3. Immediately revert to the bomb threat data sheet on the following page.

Pertinent questions to ask the caller:

1. When is the bomb going to explode?

2. Where is the bomb right now?

3. What kind of bomb is it, or, what does it look like?

4. What will cause the bomb to explode?

5. Did you or someone else place the bomb?

6. Why?

7. Who are you?

Things to remember:

1. Caller's voice, background noise, the threat (tone and language).

2. If possible call a supervisor by motion, or, if possible, by another extension.

3. Never, never break the connection of the caller; keep them on the phone as long as possible. Sometimes the call can be traced.

BOMB THREAT DATA SHEET:

Sex of caller: _____   Race:_____

Age:_____

Time of call: _____   Date:_____

Length of call:_____

CALLER'S VOICE AND DEMEANOR

Calm _____
Nasal _____
Angry _____
Excited _____
Lisp _____
Slow _____
Raspy _____
Rapid _____
Deep _____
Loud _____
Soft _____
Deep Breathing _____
Clearing Throat _____
Crying _____
Cracking Voice _____
Normal _____
Disguised _____
Distinct _____
Accent _____
Slurred _____
Familiar _____

The head of security should be notified as soon as possible. He, in turn, should notify the local FBI Office, the State Police, and the local police (in that order). While waiting for a response team to arrive, he should go over the bomb data sheet with the receiver of the call, taking notes on pertinent information, while the call is still fresh in the receiver's memory. Bomb threat calls may, or may not be real, but all should be respected as carrying the possibility of being real. Also, a bomb threat call, whether real or a hoax, is a federal offense. If you are faced with a bomb threat caller, you then must make a decision whether to evacuate the building (or premises). Evacuation will be covered in the next chapter.

7

Evacuation

WHETHER BY A THREATENING PHONE call or other distinct menace of an impending terrorist attack the credibility and source of the threat must be evaluated and a decision made as to further action must be considered. The head supervisor and the ranking security officer must make one of the following decisions based on the immediacy of the potential threat:

1. Take no action and leave it to the authorities.

2. Conduct a search without evacuation.

3. Initiate a partial evacuation.

4. Conduct a complete evacuation and search.

If evacuation is deemed necessary, you should have a primary and secondary evacuation route. These routes should be pre-planned and all employees should be familiar with both routes. Department heads, building managers or in-house security should coordinate evacuation routes awareness on-the-job and execute a leadership role on-site of the day an evacuation becomes necessary. I have noticed that many business concerns use fire alarms as evacuation signals. This is acceptable as long as the evacuees know it is an evacuation for a possible explosion and not a fire. The interior shutdown is different with an explosion than with a fire. In the case of fire, doors and windows are shut for fire containment—in the case of a possible explosion…doors and windows remain open! This is to give the pressure and thrust of the explosion an outlet and to cut down on flying debris such as glass. A "holding" area should have also been established for evacuation by supervision or security.

All evacuees should proceed to the holding area and supervisors (department heads or other lead personnel) should account for all people. The "holding" area should be at least three hundred feet from the building and away from the parking lot or parking area because of the possibility of car bombs.

If an explosion occurs, do not approach the building—just because there was one explosion doesn't mean there are no more bombs in the building, plus there is always the chance of residual damage caused by any inflammatory contents inside.

Other considerations to be weighed in deciding to evacuate:

1. The possibility of an effective search without total evacuation (not recommended).

2. Your liability if an explosion occurs and the building is not totally evacuated.

3. The proximity of adjacent buildings and the danger to them—as well as other businesses sharing the same building.

If an evacuation is deemed necessary all employees should be instructed to unlock desks and drawers, lockers, file cabinets and turn off all machinery (except lights). All purses, packages, briefcases, lunch boxes, personal property and anything else that may waste unnecessary time looking for should be left behind. If time and circumstances permit, all material that may ignite or add to fire or blast damage should be removed.

Again, doors and windows should be left open to vent and minimize blast and fragmentation damage. Once clear of the building all employees should proceed **immediately** to the "holding" area.

8

Conducting A Bomb Search

CONDUCTING A BOMB SEARCH SHOULD be left to the experts. However, there may be a time when you have no alternative and must conduct a search utilizing your own people. In that event, your security teams should conduct the search—or a contingency of supervisors trained in such matters.

Searches can be done in several different ways:

1. **Covert search:** This type of search is done quietly and surreptitiously to avoid concern or panic among employees.

2. **Open search:** This is done openly in front of the employees and employees are aware of what is going on.

3. **Evacuation search:** This is done after the building has been evacuated and is undoubtedly the most effective and safe method.

Since buildings are primary targets for any incendiary activity, the instructions for the search will be contained to a building scenario. All search team members should be aware of the different types of explosives, what appearance they may have and the different types of detonators they may encounter. Your local police department bomb squad may be willing to help with this issue by holding a class or seminar.

OUTSIDE BUILDING SEARCH

The outside of the building is an unlikely place to find explosive devise because of limited damage and casualties even if successful. However, if a terrorist finds gaining access to the inside of a building a high risk, he will utilize the outside, and you can't afford to overlook that fact (the Oklahoma City bombing). The perimeter of the building should be covered and searched in its entirety with special consideration given to bushes, shrubs and any nook, cranny or corner near or at the building's perimeter. Trashcans and all covered containers should also be checked. If possible, the outside and inside should be done simultaneously.

Searching one and then the other is too time consuming. All searchers should have knowledge of the location of all in-house phones and extensions. A CCP (Central Command Post) should also be set up. This area or room should be set up when you designate and train your search teams. So if you must conduct a search the proper personnel will know exactly where to go. The CCP is operational and ready to monitor and coordinate the search immediately. As each section of the outside perimeter is searched and cleared, the person doing the search should notify (by house phone or messenger) the CCP that "No bomb was found."

INTERIOR BUILDING SEARCH

The interior search of the building should begin with the total building first, then the room-to-room search. The search should begin from the bottom to the top. Starting with the basement area all electrical units, engine rooms, heating units, in other words, any machine, outlet, or unit that may cause residual damage if an explosion occurs.

A team should also search evacuation routes before evacuating as well. On working your way from bottom to top, after basement facilities are searched, proceed to the lobbies, restrooms, cleaning, storage closets, and elevator shafts (the top of the shaft should be searched as well).

If at all possible, and you have enough trained personnel, a "leap frog" approach should be used to save time. By a "leap frog" approach is meant two or more floors being searched simultaneously, thus allowing the searchers on one floor to skip a floor and begin the search on the next floor.

You can only do this if you have enough trained people, emphasis on **"trained"** personnel **Never** use untrained personnel in a bomb search situation. A critical point: **Never** conduct a bomb search period— unless there is no other alternative.

CONDUCTING ROOM-TO-ROOM SEARCHES

Conducting room-to-room searches should be done systematically and thoroughly. Upon entering a room, the first thing to do is, **STOP, LOOK AND LISTEN**. Look for any unusual objects, anything out of place, or, something that doesn't look quite right, listen for any unusual noise, a beeping, humming, ticking, anything making a noise that is rhythmic in nature.

The search team leader should visually divide the room according to size and structure. The room should be divided into sections according to the number of people doing the search. Once again, the search should be done from the bottom to the top. The search will consist of three passes, beginning from the sides of the room and working toward the center.

On the first pass, everything from the floor to the waist should be examined thoroughly. On the second pass, everything from the waist to the top of the head should be searched. On the third pass, everything from the top of the head to the ceiling should be examined—even acoustical ceiling panels should be lifted and checked.

Among items which should be searched are:

1. Floor coverings.

2. Room furniture.

3. Cabinets and closets.

4. Clocks and wall fixtures.

5. Sinks and laboratory facilities.

6. Loose clothing hanging or lying around.

7. Light fixtures and water coolers.

8. Trash receptacles.

9. Vending machines.

10. Public phone booths.

11. Window coverings, blinds, or drapery fixtures.

When a room has been searched thoroughly and been found clean, chalk or tape may be used to indicate that room has been searched and is clear. The CP should also be notified, by in-house phone (not cell phone or two-way radio), when a room has been cleared. Also, if you have what might be considered a "logical target," it's always a good idea to search there first. If a device or suspected device is found, **DO NOT TOUCH IT!** Note the location, description, and proximity to vital areas (e.g., gas lines, utilities, electrical facilities and call the information into the CCP). If you recognize the detonation mechanism, relay this information as well.

SEARCHING FOR CAR BOMBS

Another type of search you may never have to do is a car bomb. Even though this is routinely done in Europe and the Middle East, so far we have not had this problem in the U. S. However, this does not rule out the future of terrorism in this country and this has been a trademark of terrorists for years. As they say, an ounce of prevention is worth a pound of cure. Hopefully, we will never be faced with this type of terrorism but, by all accounts, we should be prepared for it. Somebody trained and experienced in this type of search should do a car search. Trust us on this one: A car bomb search should only be conducted by experts—

Things to do when doing a vehicle search:

1. Face the front of the car and divide it, visually, into a right half and a left half and inspect each half separately. Start at mid-front and work to the left.

2. Inspect the grill and bumper bottom, front and inside.

3. Move to the right front trim and wheel wells - this will include the bottom, front and rear of the tires where a pressure-release device could be held. Also go over the under siding of the wheel well and check the coil springs.

4. Inspect the right from front to rear tires and wheel wells.

5. Look at the trunk for signs of forced entry and also check the rear bumper for any signs of tampering.

6. Check the exhaust pipe for blockage.

7. Look at both the top and bottom of the gas tank and make sure the gas cap is locked.

8. Proceed up the left side of the car duplicating the procedures used on the right side.

CHECKING INSIDE THE VEHICLE

Before entering the car makes sure everything is as you left it. If you left it locked, make sure it is still locked (if you find the vehicle unlocked, report the incident and do not enter vehicle; remember, **EXPERTS ONLY**). Check the weather stripping around the doors and windows for tampering. If everything is clear then unlock and enter the vehicle.

Proceed to inspect the following accessible points of:

1. Driver and passenger seat.

2. Headrest.

3. Rear seats.

4. Dashboard.

5. Glove compartment.

6. Sun visors.

7. Alarm system.

8. Finally, release the hood latch and inspect the engine.

SEARCH TEAM SUPERVISOR

It is essential that the activities of a search team be coordinated efficiently, professionally and thoroughly. The supervisor should work from a CCP and be in constant contact with his search team members. All activities of the search team should be relayed to the Supervisor by telephone or in person. He/or she needs to coordinate the activities of his team in a timely and effective manner.

The supervisor should be thoroughly trained, a quick thinking and capable of making intelligent decisions under stressful conditions. It cannot be emphasized enough that doing a bomb search should be your last resort and done only if a possible explosion can do irreparable damage and the explosive must be found at all costs and there are no other alternatives. If you must do a search, your function is to only locate the device and then vacate. Call the FBI, your state police and the local authorities.

If an explosion does occur, **DO NOT TAMPER WITH DEBRIS**. Call the appropriate authorities, care for any casualties, have security cordon off the area from spectators and media and keep the area clear for ambulance and authorities. You may want to advise anyone in adjacent buildings to evacuate because a blast may loosen or weaken structures in the surrounding area. Exercise all caution in avoiding post-explosion catastrophes.

9

Anti-terrorist Security—The Human Factor

LIGHTS, CAMERAS, ALARMS AND OTHER means of technology can aid in discovering and deterring terrorist activity but the end result lies in the hands of the men/women who, monitor, watch and act on what they can get from all of the anti-terrorist means we have initiated. Things like cameras and lighting can provide us a means to an end, but ultimately, unless we utilize what the lights and cameras can show us, we are still at risk.

This brings us to the subject of the adequacy of your security personnel. In the following chapter I will offer my opinion from personal observation about private security—heed this well.

Things you need to keep in mind in evaluating your security:

1. Are they alert to their surroundings?

2. Are they certified in different areas (baton, pepper spray, CPR, first aid, firearms)?

3. What is the extent of their training?

4. Do they know how to recognize an incendiary devise?

45

5. Do they perform their rounds professionally or perfunctory (checking doors, checking containers, etc.)?

6. Have they been trained in crowd control and evacuation procedures?

7. Are they familiar with evidence gathering?

8. Do they know how to properly search a vehicle?

9. Are they efficient in writing reports and documentation?

10. Are they professional in appearance and demeanor?

These are all important in having an efficient anti-terrorist force in place. It's almost mandatory in today's climate to use armed security. Unarmed security is a thing of the past (no one respects only uniforms anymore). If you are at risk from a terrorist attack, you not only put yourself in jeopardy using unarmed security; you risk your security guard's life as well. I would, in fact, if you were a large enough company and concerned with a terrorist attack, develop your own security program. To not do so is to invite disaster.

Use people that have a loyalty and commitment to your company, those that work for you and you alone and have a vested interest in your company and your employees' well being. By hiring an outside source, a source who may work for you and other companies, you are taking a chance on your safety by leaving it in the hands of people who do not have your interest as a primary concern—the security company wants a contract and the security guard just wants paid—end of story. It will cost to develop your own security, but in the long run it will pay for itself.

You will not only have security guards but also individuals who work for you, who are paid by you and are company employees (yes, full-time insurance benefits; don't skimp). This will be a dedicated, loyal, worker whose livelihood and future lies in your success and safety. This will also help stop the turnover rate (which is rampant in private security) and you directly control your own security and personnel without having to go to an outside security company. Eventually, I can see most companies going down this path because using private security, the way it is today, is risky, dangerous, and you don't know what you are getting. An excellent example of a company developing their own security, that I have had the pleasure to work with happened in 1996. I was Supervisor-in-Charge of a group of ex-military personnel, working at the CNN Building during the 1996 Olympic Games. Our mission was to conduct Anti-terrorist Security in support of Turner Security. CNN has it's own security and they are the most professional, best-trained security with whom I have ever worked. They are motivated **because** they are part of CNN. When the bomb went off in Centennial Park, security at CNN became extremely tight. Turner Security reacted in a very professional and timely manner. The experience at the 1996 Olympics was very stressful but it showed (me) how good an armed security team could perform if well trained, well paid and worked for the company they protected.

On the other hand, let's look at the protection of our nuclear facilities. Rich Lavernier, who runs mock-terrorist raids on nuclear facilities using Special Forces, who was interviewed by Sixty Minutes, said, "We have a fifty per cent success rate in penetrating nuclear weapons facilities." His findings were more or less dismissed by the Nuclear Security Agency (NSA), by saying "...we are working on the problems." As a result of his findings, Lavernier was demoted.

Security was even told the date that the mock penetration would occur. On a "pop" security check, they found patrols were absent. Most of the security personnel were in one central place watching the Super Bowl. They also found guards sleeping, a definite no-no in the security business—but universal in the private security sector.

Another problem they found was the loss of card keys. Codes were not changed as a result of the "loss," so anybody finding (or stealing one) had immediate access to facilities. If this doesn't scare you, it should—it scares the hell out of me. Such is the security protecting our most formidable weapons. Ask yourself the telling question, especially if you are the CEO in charge of your company, "Who is protecting your future?"

10

Family Guidelines

IT WOULD BE REMISS OF us not to include a chapter on "the family" and their safety. Even though this has been written to provide business and industry some insight into anti-terrorist security we cannot overlook the fact that the family unit is the heart and soul of business' and industries. As responsible mothers, fathers, grandparents, and siblings, our first duty is to our family, to protect, provide for, and make sure they remain safe if a terrorist act does occur. The Department of Homeland Security has been very active in making public service announcements on radio.

There are also many organizations from which you can obtain helpful guides on emergency planning. Through the years the American public has lived a peaceful and complacent life. War and terror were the problems of countries abroad and our lives were only concerned with day-to-day living and reading about war and terror elsewhere through the media. Now, our lives have changed, and the one, single factor that must be included in our lives today is, **awareness**—awareness of our surroundings, work place, when going and coming from work, even while shopping.

We are not used to doing this and it is very difficult to re-train ourselves to get into that mind set. Being prepared is not just a job for professionals (law enforcement, firefighters, federal agencies, etc.). All of us need to

49

be able to prepare for an emergency and react instinctively during and after a terrorist attack. One of Homeland Security's main goals is to continually educate the public on emergency preparedness, including the possibility of terrorist attack.

Homeland Security recommends an initial three step plan that families can take:

1. Assemble an emergency supply kit.

2. Develop a family communication plan.

3. Learn more about the different types of emergencies with which you could be faced.

EMERGENCY SUPPLY KITS

In putting together an emergency kit, it is first recommend that you pick one room in your house or apartment as a central storage place— where everyone in your house knows things are located. It should be a room that has the least number of windows, preferably no windows. A disaster of any kind may interfere with normal supplies of food, water, heat and normal day-to-day necessities.

It is important to keep a stock of emergency supplies on hand sufficient to meet your families needs for a least three-days. Although a natural disaster or terrorist event may not directly impact you in your area, utilities and other services may be affected. In an emergency event it may take sometime to restore power or other services to you and your family. That is why an emergency supply kit is essential so that you and your family can take care of yourselves for a three-day period of time. Having the necessary supplies will ensure that you are cared for.

If you are forced to remain in your house during a natural disaster or terrorist event, adequate supplies could help you survive through a period of danger without hardship. If you are directed to evacuate instead of sheltering in a place, an emergency supply kit can be taken with you and used in a shelter. You should prepare two kits and store supplies in watertight containers.

Keep one easily accessible in your home and one in the trunk of your car. It is advisable that you update your kit regularly. You should replace the water supply and any food that may have reached its use-by or expiration date.

An easy way to remind oneself to update your kit is to use Daylight Savings Time. When you change your clocks, you should also check your kits. It is also a good time to check the batteries in your home smoke alarms and other household safety equipment.

An emergency supply kit should include the following:

1. Bottled drinking water—one gallon per day per person with at least a three-day supply for each person in your household. Other drinkable liquids (fruit and vegetable juices, soft drinks, etc.) can be stocked.

2. A battery powered radio and a flashlight, with extra batteries for each.

3. Canned or sealed packaged foods that do not require refrigeration or cooking.

4. Blankets or sleeping bags for each member of the family.

5. Writing materials to take notes from television or radio broadcasts.

6. Bandages.

7. A can opener.

8. Fire extinguisher.

9. Signal flare and whistle.

10. Cell phone.

11. Soap and towels.

12. Paper towels and toilet paper.

13. Household laundry bleach (unscented).

14. An extra set of car keys and a credit card, cash or traveler's checks.

15. One change of clothing and footwear per person.

16. First-aid kit and manual.

17. Non-prescription drugs, mild pain relievers and antiseptic.

18. Any special prescription medicines or foods needed by family members such as insulin heart medication, dietetic food and baby food (do not store in the kit for long periods of time and rotate out every so often).

19. Extra eye glasses, contact lenses and supplies.

20. Games or books

21. You also can store additional water by filling bathtubs and sinks with water if an emergency is declared. Clean water is also available in toilet tanks, presuming that no chemicals and other cleaning agents are not used in the water tank.

Additional Supplies Items:

1. A cooler, freezer packs, and ice.

2. Salt and pepper.

3. Paper cups.

4. Baking soda.

5. Spoons and forks.

6. Matches or lighters.

7. Tape.

8. Needles and thread.

9. Pre-moistened towelettes.

10. Hand lotion.

11. Disposable diapers (if you have infants).

12. Cotton balls.

13. Sanitary napkins.

11

Two Views Concerning Private Security

THIS CHAPTER DETAILS TWO PERSPECTIVES of private security. The first is one I originally wrote (A) in November 2001, right after September 11. I had worked with private security in many states and had to be licensed in a number of those states. The ease of getting a private security license, the lack of practical training and the quality of individuals I found working in private security has, since September 11, made private security a liability. Having tried to change this inequity in the state of Virginia and wasting time and effort talking to politicians who didn't care about the state of security, either locally or nationally, I simply gave up further efforts to do so. As I mentioned in the introduction, I was interviewed by "Sixty Minutes" who had received a copy of this brief and were planning a segment on private security. The second assessment of private security was the front-page story (B) in "USA Today," January 23, 2003, fourteen months after I had written mine. The following are both articles printed word for word.

THE STATE OF PRIVATE SECURITY

By: Loren Jackson

I have been involved in the field of security for the past ten years. I do not work in the field of private security, but rather, in the field of strike security and executive protection, which is a little more volatile and hazardous than regular private security.

However, I have worked in twenty-one states and had to be licensed in twelve of these states. Besides that, I have had to work with private security personnel in almost all of my assignments.

What I have noticed, first of all, with the licensing procedure, is that all states I have had to be licensed in is the similarity in styles of training and the total lack of practical training involved. Licensing procedures were very similar in each state—a class lasting from eight to sixteen hours (depending on the state). In this classroom a prospective security guard is basically taught what he can, or can't do, as a security guard in that particular state.

He is taught the meaning of various legal terms, and, in some states are versed in rudimentary first aid, and how to tell the difference between the various fire extinguishers and what fires to use them on. After the mandatory classroom requirements are met, the potential licensee is given a test. If he passes the test he is then licensed.

Most states do have a state or federal background check (however thirteen do not), that the security guard must pass. What is lacking in training is the absence of any practical training that is useful in security (especially at this point in time).

They are not taught what to do in emergency situations, crowd control, CPR, explosive recognition, fire prevention, and evacuation procedures. My whole point here is that you can get a security license easier, and with less effort, than a sixteen year old can get a driver's license.

Private security is full of ill trained, unqualified, low-paid, non-motivated, uncaring, individuals who are supposed to mirror the image of the professional security guard. In reality, they are simply bodies occupying time and space with no apparent concept of what their job is how to do it properly, or what security is all about.

Security companies are forced to take anyone breathing, who holds a security license (and believe me, anyone breathing can get a security license), and put them on a job site without a proper job description, simply to fulfill their security contract. Private security is a joke and most people realize it, but no one will address it. I have worked with private security officers all over the country and they are the same everywhere.

Many of them sleep on duty, some have medical problems that should exempt them from most jobs, but security will hire them, many are so irresponsible they don't even show up for their job and don't even bother calling to let their supervisor know they aren't coming. This is not second hand information—this is what I have actually seen.

We seem to have two types of people working private security—the elderly, supplementing social security or retirement, or the young, who are simply working for a check until they find a better paying job. The older security guards are the most professional in actions and attitudes but they are not trained and physical limitations may prevent them from being effective in emergency situations.

The younger security guards are usually irresponsible, uncaring, hate the job, and will sleep if they can get away with it and these are the ones who simply do not show up for duty or quit without notice. What needs to be done with private security is a complete revamping to appeal to the more responsible individual. These are the middle-aged people who have a family and responsibilities, who take their job seriously and strive to better themselves.

Until we reform security to appeal to this kind of person, we are going to be stuck with what we have. Private security, which has been regarded with everything from amusement to contempt, has now become a liability, and something must be done.

By using the security personnel we have now we are putting, not only the lives and property of the company he is doing security for, we are putting the life of that security guard in danger.

We, in the Tidewater Region, are in an especially high-risk area and should act accordingly. We have tunnels, naval yards, shipbuilding, military bases, and even a nuclear facility close nearby.

We need to turn private security from a negative to a positive in Virginia. Chuck Vance, founder of Vance International, in his newsletter, pointed out that private security outnumbers law enforcement five to one.

We have an anti-terrorist force in place, why not train, screen, and pay them accordingly? This would be the most cost effective and practical way to combat terrorism.

In order to change the state and nature of private security, and implement it to support and supplement law enforcement, I humbly submit the following recommendations:

1. Revise the training, testing, and licensing procedures in Virginia.

2. Appeal to better qualified more responsible people.

3. Put an end to bidding wars where companies accede to the lowest bidder (lessons from airport security should have shown us that).

4. Make private security a **PROFESSION**, not a job.

REVISE TRAINING, TESTING AND LICENSING

Training should include classroom requirements but far more in depth than what is done now. Before September 11, a security officer's main function was to observe and report. Today, however, he must also be taught how to react to any given situation. His training should include first-aid, CPR, fire control and prevention, crowd control, explosive recognition, how to handle a bomb threat call, how to conduct a bomb search, and, self-defense techniques.

I also feel strongly that unarmed security welcomes terrorist acts and feel that security guards be armed certified, baton certified and certified in the use of pepper spray. I think this training program should be regulated by the state and use the facilities of the state police and/or the local police training facilities.

State and local police officers can also serve as instructors. I was sent to a training school for a week and versed in all of the above by a private company. If a private company can do this surely a state can do it. The classroom portion should be a minimum of sixty hours and the practical training course at least a week. At the end of this section I have included a prototype of what the training course could consist of.

APPEAL TO BETTER QUALIFIED, MORE RESPONSIBLE PEOPLE

Private security is made up of a diversity of people who work in security for a number of reasons, "moonlighting" to supplement social security, retirement or another job.

Working to pay for school, or until they can find another, better paying job, and, unfortunately, people who can't find work anywhere else. Whatever the reason, if it is any of the reasons I have just mentioned— **IT IS FOR THE WRONG REASON!** Money, however menial, is their primary motive.

Right now private security is full of unqualified, irresponsible people. Many of them are unreliable and work only when they want (they know they won't be fired because the company needs bodies).

A security company's primary concern is, will all posts be manned? Their concern is making sure all posts are manned to insure their keeping the security contract—security isn't even on their agenda. Yet today, when security is a priority, we put our trust in people and security companies where security takes a back seat to their own concerns.

We must change the very nature of private security and to do this we must appeal to a better-qualified and more responsible individual. To do this the pay structure, the chance for upward mobility, and the possibility of a career must be implemented.

The future of this country will depend on the strength of our security and we **MUST** reform it to appeal to people who are responsible, motivated, talented, and ambitious. It seems we would wake up to the changes that need to be made but we haven't. Will it take another terrorist act with lives lost on a major scale for us to wake up?

THE LOWEST BIDDER

As we all learned after September 11, security contracts invariably go to the lowest bidder. Now, it becomes a question of, "How much value do you put on your life and property?" The competition in security is so great that companies underbid to get contracts. Consequently, they pay security guards relatively little; they can only attract those desperate for work. This is why security is in such a mess now, low bidding, hence, low paying. So your security is second rate at the very least. Yet we are faced with the most dangerous enemy we have ever had, and on our soil.

All of our security is in the hands of a security guard making barely more than minimum wage, untrained, to him we entrust our lives, property and safety, while facing a formidable foe that even soldiers have a hard time dealing with.

This just doesn't make sense. Somehow a ceiling should be put on this low bidding concept so security companies can provide a decent wage for their guards, provide benefits and thus, appeal to a higher caliber person. Janitorial services also rely on low bidding to secure jobs—I, for one, hate to think my safety, the safety of my children, and the safety of the nation dwells in the same realm as the janitorial business. Again, I would like to quote Chuck Vance; "We must rely on the eyes, ears, and deterrent effect of a private security force. How can we, with any type of certainty, do this if we do not provide these people with a livable wage, benefits, training, equipment, and proper supervision? The bar must be raised on the quality of our private security forces, not for the "knee jerk" short term, but from this day forward."

MAKE PRIVATE SECURITY A PROFESSION

Hopefully, September 11 will be etched in our memories forever. We should, at this point, be reevaluating our position and considering measures to keep this from happening again. The single, most effective weapon we have is **SECURITY**. Not the security we have known in the past or as we know it today. But, a security force that is administered to professionally, that is made up of professional, qualified, trained, career-minded security personnel. We take pride in our law enforcement institutions, just as they take pride in themselves.

We must adjust private security where the people who serve take pride in what they do and make security a profession, not just a job. We must have a security force, well trained, well paid, proud of their profession, and dedicated to preserving and protecting the people and property of the Commonwealth of Virginia. Security must become a professional institution, with professional people, doing a professional job.

If we don't start changing, and changing now, we are endangering everything and everyone. Private security is a liability, not an asset. The people of Virginia have elected you to represent us and we depend on you for our safety and welfare.

As private citizens, we can do nothing to improve private security except appeal to you, our elected officials. We depend on private security to protect our buildings, our banks, our post offices, our schools, sporting events, concerts, and other things too numerous to mention.

With private security playing that big a role in our lives, and now that terrorism has become a clear and present danger, security is no longer an issue—it is a **PRIORITY!**

The following is a training schedule prototype. As mentioned earlier, we could use the barracks and training facilities of local and state law enforcement for training purposes and even use some of their personnel as instructors in the classroom and in the training arena.

Prototype of Training Schedule

Sunday:

Arrive by **5:00PM**

5:00PM – 5:45PM—Light Dinner

6:00PM – 7:00PM—Orientation and Schedule for the week

Monday:

6:00AM – 6:45AM—Physical Training

7:00AM – 7:45AM—Breakfast

8:00AM – 12:00AM—Standard Security Class

12:00AM – 12:45PM—Lunch

1:00PM – 5:00PM—Standard Class Continued

5:00PM – 5:45PM—Dinner

6:00PM – 7:00PM—Review

Tuesday:

6:00AM – 6:45AM—Physical Training

7:00AM – 7:45AM—Breakfast

8:00AM – 12:00AM—Explosive Recognition (Types, initiators, power sources)

12:00AM – 12:45PM—Lunch

1:00PM – 5:00PM—OC Certification

5:00PM – 5:45PM—Dinner

6:00PM – 7:00PM—Review

Wednesday:

6:00AM – 6:45AM—Physical Training

7:00AM – 7:45AM—Breakfast

8:00AM – 12:00AM—Baton Certification

12:00AM – 12:45PM—Lunch

1:00PM – 5:00PM—Baton Training Continued

5:00PM – 5:45PM—Dinner

6:00PM – 7:00PM—Review

Thursday:

6:00AM – 6:45AM—Physical Training

7:00AM – 7:45AM—Breakfast

8:00AM – 12:00AM—Firearm Certification

12:00AM – 12:45PM—Lunch

1:00PM – 5:00PM—Firearm Training Continued

5:00PM – 5:45PM—Dinner

6:00PM – 7:00PM—Review

Friday:

6:00AM – 6:45AM—Physical Training

7:00AM – 7:45AM—Breakfast

8:00AM – 12:00AM—First Aid

12:00AM – 12:45PM—Lunch

1:00PM – 5:00PM—Crowd Control

5:00PM – 5:45PM—Dinner

6:00PM – 7:00PM—Review

Saturday:

6:00AM – 6:45AM – Physical Training

7:00AM – 7:45AM—Breakfast

8:00AM – 12:00AM – Fire Fighting and Prevention

12:00AM – 12:45PM – Lunch

1:00PM – 5:00PM – Basic Self Defense

5:00PM – 5:45PM – Dinner

6:00PM – 7:00PM – Review

<u>Sunday:</u>

6:00AM – 6:45AM – Physical Training

7:00AM – 7:45AM—Breakfast

8:00AM – 10:00AM – Awareness – What to look for

10:00AM – 12:00PM – Review

12:00PM – 12:45PM – Lunch

1:00PM – 3:00PM – Exam

3:00PM – Dismissed

Again, the above was an assessment I wrote in November of 2001. The following article appeared as the front-page headline in USA Today, on January 23, 2003. There are many similarities but the USA Today report has a lot of research material that I lacked. The important thing to remember is my report on private security was based on personal experience; the report in USA Today was based on pure research. The frightening fact is both reports reinforce each other in analysis and content.

Private security guards are Homeland's weak link

By Mimi Hall, USA TODAY

(States Requirements Table follows)

They are the first lines of defense against terrorists. But more often than not, private security guards that protect millions of lives and billions of dollars in real estate offer a false sense of security. Most of the nation's 1,000,000 million-plus guards are unlicensed, untrained and not subject to background checks. They're burgeoning, high turnover, low pay, few benefits and scant oversight mark $12 billion-a-year industry. According to government officials and industry experts, little has changed since September 11, 2001. As the demand for security guards increases, security companies "find someone on the street and put him in a uniform. Before he or she is finished buttoning up, they put him or her on a post," says Henry Nocella, vice president of Professional Security Bureau, a private company based in New Jersey that employs about 4,000 guards. Thus the name: "rent-a-cops."

For 16 months since terrorists toppled the World Trade Center towers and destroyed part of the Pentagon, government officials have worked to secure the homeland. But there are no federal laws governing the private security industry. Efforts in Congress to mandate training and background checks nationwide failed last year; sponsors expect better results this year.

State laws remain spotty. While the tiny fraction of guards who carry guns go through training and background checks, most of those who patrol office buildings, apartment complexes, shopping malls, sports arenas, warehouses and cargo terminals are unarmed and technically untrained. Twenty-nine states and the District of Columbia do not require training for unarmed guards. In 22 states, they don't have to be licensed. In 16 states, the people who hold the keys have access to the ventilation systems and know the escape routes are not put through background checks.

Security guards themselves say they have seen few improvements since the 2001 attacks. A poll of 1,200 guards in California, Texas and Florida last spring for the Service Employees International Union found lax security persisted. Four in 10 guards said their buildings had no new security procedures. Seven in 10 said no bomb-threat or natural-disaster drills were conducted. A majority said they received no training on evacuation or other emergency procedures before being hired.

"Post-9/11 training is non-existent in certain parts of the country," says Bruce Gelting of Allied Security, the largest American-owned and operated Security Company with 19,000 guards.

Many large companies hire top-flight security firms. They teach guards how to spot suspicious packages, monitor security cameras, and evacuate buildings. But there are at least 11,000 security firms nationwide. Experts say many are fly-by-night operations that invest little or nothing in training and don't check backgrounds.

Many Fortune 500 companies "are just putting bodies in uniforms," Gelting says. "Not trained for terrorism. Raynard Williams is an $8.40 an hour guard at ABC Entertainment Center, an entertainment complex in Los Angeles. To get his job with Universal Protection Service (UPS), he sat through a generic, four-hour training video." "Something to put you to sleep," Williams said. Like most guards, Williams, 39, gets no health insurance through his job, no paid vacation and no sick days. "It's a thankless job," he went on to say.

In 2000, the most recent year for which figures exist, private security guards earned an average of $17,570. For many, it's a second job; most leave within months. Williams says he and his fellow guards feel vulnerable without proper training. But employers, he says, know that "if they give us training, we'll want more money." That's a big reason for the long-standing opposition to government regulation. In an industry in which contracts are awarded to the low bidder, private security companies oppose government mandates that would increase operating costs. The businesses that hire them for protection don't want those costs passed along.

"The security guard industry is a very competitive one, and their contracts are won and lost based on pennies per hour," says Jeff Schlanger of the risk consulting company Kroll, based in New York. "It's all about money." The point being made here is clear enough—even after the horror of September 11, analysts say, most companies are reluctant to pay more for security, regardless. In a tight economy, companies are "Looking right down the gun barrel of some tough economic issues. What does it cost and what do we get for it?" says Philip LaRiviere, a security consultant based in Chicago. Companies are buying more security, but they still hire the low bidder.

It's hard for security companies to prove their value. Gail Simonton of the National Association of Security Companies, which represents the nation's largest outfits, says they can't prove that hiring guards will avert disasters. "You're trying to prove a negative," she says.

Experts say if the government doesn't demand higher standards, the industry will continue to provide a dangerous opportunity for terrorists. Some could slip by untrained guards. In other cases, would-be terrorists could infiltrate the system by getting work as guards themselves.

Over the years, criminals have landed jobs as security guards. Some didn't go through background checks. Others were subject only to one state's checks, which didn't find criminal records in other states. Stories of guards beating, raping and robbing the people they were hired to protect have hurt the industry's image.

Those problems persist. The New York Daily News reported last summer that security companies hired by the state to protect the Statue of Liberty and other state and military facilities employed hundreds of unlicensed guards, including former convicts.

In Atlanta last spring, federal investigators found that private security guards employed by the federal government to protect four federal buildings were easily duped by undercover investigators! The investigators were able to talk their way through security without identification and slip weapons into the buildings. In one security breach, an investigator who entered a building with no ID persuaded a security guard to give him a pass and a special access code to enter the building at night.

Janet Boston knows firsthand the chaos that can ensue when untrained guards face an emergency. A guard at the World Trade Center for 26 years, she was on the 78th Floor when terrorists set off a bomb in 1993 at garage level. She and other security officers were as clueless about what to do as workers hired to protect the building and its people. They didn't know whether to evacuate the building. They didn't know where to tell panicked people to go. As a result, workers rushed down stairwells and were blocked by locked fire doors.

"People were hollering and screaming all over the place," Boston recalls. "Nobody knew what to do, not even security. It took almost the whole day for us to figure out what was going on. People were hysterical." After the 1993 incident, the Trade Center's guards were given extensive training. It included a 40-hour course that taught them which floors were blocked by fire doors and how to evacuate thousands of workers in orderly fashion. Training updates and drills continued monthly.

Experts say the guards at the World Trade Center became the best trained in the country. On September 11, "that training saved thousands of lives," Schlanger says. Security guards helped guide thousands of workers to safety before the towers fell. Since September 11, there have been no catastrophic events in the United States to test the quality of the private sector security force. But government officials warn that terrorists will strike again, and experts say the private security industry's lack of standards leaves millions vulnerable. Security guards "are the ones who are going to notice that something's out of place," Nocella says. "If they don't know what steps to take, it's not going to matter how many firefighters or police or first responders you have because it's going to be over." In countries such as England and Israel, where terrorist attacks have been a way of life for decades, security is treated more seriously. But in the United States, with the exception of the armed and highly trained guards who protect federally regulated nuclear plants, "we run much more a hotel concierge version of security," says Andrew Stern, head of the Service Employees International Union.

'Congress ought to act'

Among unions, employers and elected officials in both political parties, there is a growing movement to improve standards and regulate the fast-growing private security industry.

Sen. Carl Levin, D-Mich., is pushing a bill that would give states easy access to the FBI's database for criminal background checks of potential guards. That's intended to make sure terrorists doesn't slip into the system." It's a real vulnerability that we put such heavy reliance on the private guard service but do not have guards who are well trained or put through checks," Levin says. He notes that Congress has passed laws giving nursing homes, day care centers and banks access to the database.

A similar bill in the House of Representatives also would give states financial incentives to require 40 hours of training before putting guards on posts. Last year, the bill was co-sponsored by a liberal Democrat, Rep. Dennis Kucinich of Ohio, and a conservative Republican, Rep. Bob Barr of Georgia, who has since left Congress. Supporters say they will push for passage again this year. "Congress ought to act because the private security forces are part of the homeland security strategy of the future," says Robert McCrie, a professor at John Jay College of Criminal Justice in New York.

Some states have started to act. Former California Gov. Gray Davis signed legislation before leaving office that required guards to get 40 hours of training—eight before being put on a post and 32 more within six months. California also is among 22 states that require each guard be put through a federal background check. It would pick up a criminal record anywhere in the country, not just in the state where the guard applies for work.

Most industry experts agree that federal background checks should be conducted for guards nationwide. But they disagree about whether the federal government should mandate a certain number of training hours.

Allied's Gelting says his company's guards receive site-specific training tailored to their posts. He opposes a federal mandate for 40 hours of training because "someone guarding a warehouse in the middle of Iowa is a lot different from someone at a New York City warehouse." Many industry experts share that concern.

Meanwhile, the industry continues to grow with the demand for more guards. The Labor Department predicts that employment of security guards is likely to grow faster than the average for all occupations through 2010. The main reason concerns terrorism.

Experts say that's all the more reason for better standards and training. "If history proves itself out, we will have some day a huge disaster at a building again," Stern says. "And as opposed to the World Trade Center, where we saw the value of a well-trained force, we will have excessive deaths."

(State Requirements Table follows)

Comparing local requirements

Most of the 1,000,000 million-plus private sector security guards lack training and are not subjected to background checks.

Rules for Washington, D. C. and the following states:

State	Hours of training required	Background check type
Ala.	None	None
Alaska	48	None
Ariz.	8	Federal
Ark.	6	Federal
Calif.	40	Federal
Colo.	None	None
Conn.	None	Federal
Del.	None	Federal

D.C.	None	Federal
Fla.	40	Federal
Ga.	8	Federal
Hawaii	None	None
Idaho	None	None
Ill.	20	Federal
Ind.	None	None
Iowa	None	Federal
Kan.	None	None
Ky.	None	None
La.	16	State
Maine	None	State
Md.	None	State
Mass.	None	None
Mich.	None	State
Minn.	12	Federal
Miss.	None	None
Mo.	None	None
Mont.	None	Federal
Neb.	None	None
Nev.	4	Federal
N.H.	None	State
N.J.	None	Federal
N.M.	None	State
N.Y.	24	State
N.C.	4	Federal
N.D.	32	Federal
Ohio	None	State
Okla.	40	Federal
Ore.	12	Federal
Pa.	None	State
R.I.	None	None

S.C.	4	State
S.D.	None	None
Tenn.	4	Federal
Texas	1	Federal
Utah	8	Federal
Vt.	None	Federal
Va.	16	State
Wash.	4	State
W Va.	None	None
Wis.	None	Federal
Wyo.	None	None

12

Alternatives

SEPTEMBER 11 HAS CHANGED DRAMATICALLY the lives of all Americans. It took away the innocence of complacency and demanded a response of retribution and public awareness. We have been to war in Afghanistan and Iraq and have given notice to terrorists worldwide. But on the home front we have shown we are vulnerable and must regroup and prepare to defend our country, our liberties and our very lives. That's the kind of war we are in, now and tomorrow. The Moslem terrorists have had years to build networks within the U. S. under the safeguard of our very constitution and now we have to weed them out with whatever means necessary. In the meantime, we have to prepare to face what the future may bring which means revamping all of our security systems—especially private sector anti-terrorist security.

In the realm of private sector security, all of our emphasis has been in deterring criminal activity. All of our checkpoints, guard shacks, cameras and entire perspective have been built around the criminal element. Terrorism is a completely different animal—far more violent in nature than the normal criminal we face. Terrorists' intentions are far broader in scope than those of a criminal, his actions far more fanatical. Another big difference between the criminal mind and that of a terrorist is that a criminal wants to get in a secured place, then perform his criminal act (e.g., stealing, theft, robbery) and then get out safely. A terrorist only wants in, as if he penetrates security someone dies— At that point, the terrorist doesn't care if he gets out. So, our primary goal has to be keeping the terrorist "out" with iron clad security measures.

Right now, we do not have that—we have security searching vehicles not knowing what they are looking for, we have security guarding our public buildings, our businesses, industry and none of them have had the least bit of anti-terrorist training. Airport security has changed for the good, more thorough and proficient and this is evident to all. But this has been through the efforts of the Federal Government. The same thing must be done in the private sector, yet none of the states have been willing to set a precedent and change security in the private sector realm.

Apparently, the national inactivity sits waiting until another terrorist act takes more American lives and causes more damage until the "powers that be" wake up and do something. Until that time, we are on our own in providing adequate security for our companies, employees, and our own safety. We have to do something within our capacity to provide security and make ourselves a "hard target."

Your alternatives are, basically, very simple:

1. You can stay with the status quo and hope we do not get hit.

2. You can carry on with what security we have, regardless of their ineptness and lack of training.

3. You can use this manual to help solidify your security as best you can.

4. You can have a company come in and do an anti-terrorist survey, making recommendations on how you can better secure your business.

5. You can have a company come in and do a turnkey operation, training everyone from employees to your own company security.

Regardless of the decision you make to protect yourself from terrorism, we all must be aware of the willingness of terrorist groups to strike at any target large or small that has vulnerability and weak defenses. In the past terrorist threats and actions have been aimed at government representatives, high-profile figures and political heads. This is no longer true, they will strike anywhere and there are no innocents—we are all at risk and must take this threat seriously.

Most experts say that terrorists are planning a massive plot that will be more horrendous and destructive than September 11. This is probably true, and their historic trademark and methods of operation will not change. They will still target the private sector, carrying out as much death and destruction as they possibly can. This is what we must prepare for and guard against and expect down the road.

I believe it may be years before the arms of Homeland Security will be felt within the private sector. They are concentrating on sealing the borders of the U. S. and rightfully so, to keep known or suspected terrorists outside, no longer able to hide behind the Constitution of the United States and our Bill of Rights.

The FBI is overworked and undermanned for the task they have before them, which is rounding up possible terrorists, uncovering cells and plots, following possible leads and at the same time carrying out normal criminal pursuits. It is a seemingly impossible task and leaves very little leeway in advising and aiding the private sector.

It is up to us, citizens all, to provide for ourselves until priorities are taken care of by federal agencies across the board. In the next chapter the workings and priorities of the "Knight's O' the Round Table" are explained in detail, along with their history and agenda.

13

The Knights O' the Round Table

SCOTT KNIGHT AND RON GAUNT founded **The Knight's O' the Round Table** in February 2000. At that time, the company's primary function was doing Strike Security, Executive Protection, and Private Sector Security.

Starting in the coalfields of West Virginia they expanded their operations and have worked cities from New York to Los Angeles, establishing their reputation as a professional, highly competent force with expertly trained and respected individuals. Since September 11, however, their priorities have changed. Seeing a void in the business sector that the federal government could not immediately fill, they restructured their capabilities and capacities into an Anti-terrorist Security Force.

Working in the business and industrial sector, they provide expertise and instruction in Anti-terrorist Private Sector Security. Working under Action Enterprises, an LLC based in Virginia, **The Knight's O' the Round Table** furnishes the private sector with a broad scope of services geared to deter and prevent terrorist activities. They offer anti-terrorist surveys to point out flaws in your present security and recommendations for improving security measures, including premises lighting, surveillance cameras, entrance procedures, present security personnel evaluation, overall operating procedure security and any other recommendations that deal with the safety and well-being of your employees and property.

Surveys take from three days to two weeks depending on the size of the facility and nature of the survey. The survey encompasses both day and night reconnaissance of the premises and the recommendations are based on the needs and re-structuring of your security to make you a "hard target" for any terrorist activity. The survey will not, in any way, impede nor interfere with normal operating procedures. At the end of the survey a meeting of the principals is scheduled and all points are covered in detail. A written summary is also provided for future reference.

In doing several surveys myself, I have found that private industry is lacking in both security procedures and physical deterrents to terrorist activity. A survey is only one of the functions provided by the **The Knight's O' the Round Table**. Using experts in the field, they provide instructors in baton certification, pepper spray certification, armed certification, self-defense and explosive recognition for your security personnel. They also can train your receptionist in handling bomb threat calls, provide evacuation routes, CPR certification, first-aid instruction, computer security, provide K-9 training and provide explosive-sensitive dogs. Doing a turnkey operation includes bringing the executive staff, receptionists, maintenance and security into play as one organized unit.

Again, depending on the size of the facility and the scope of the job to be done will dictate the length of time to do the survey. They will make themselves available as long as the client needs them and also be available for follow-up training. They also have the capacity to hire and train your own security force if you wish to form your own security team. Their training will entail much more than your state requires and they will be thoroughly indoctrinated in anti-terrorist techniques and security. Either supplying the K-9 units, or training your own K-9 detail can also be utilized.

The entire training program is unique because it covers every phase of anti-terrorism, from a briefing on the Islam religion to using practical methods of training and testing of all pertinent personnel within the company and the entire security staff.

A turnkey operation consists of the following:

Executive Staff:

1. Orientation and operation procedure.

2. Appoint and train evacuation chairman.

3. Find and equip situation room for emergencies.

4. Locate and check all fire extinguishers and first aid stations.

5. Include head of maintenance and safety in all of these situations.

6. Plan evacuation and secondary evacuation route.

7. Locate staging area and secondary area.

Receptionist(s):

1. Go over bomb threat data sheet.

2. Instruct in how to remain calm in case of threatening call.

3. Test with mock call.

Security:

1. Orientation:

    a)   Nature and history of terrorism.

b)   Importance of appearance and demeanor in security.

c)   What to look for.

d)   Observing, reporting, reacting.

e)   Baton Certification.

f)   OC (pepper spray certification).

g)   Armed Certification (according to state regulations only if client requests certification).

2. How to conduct a vehicle search.

3. How to conduct an inside bomb search.

4. How to conduct a car bomb search.

5. How to handle a bomb threat call.

6. Basic self-defense.

7. Emergency tactics.

8. Evacuation.

9. Basic first aid.

10. CPR.

11. Fire prevention and fighting.

12. Explosive recognition.

13. What to look for as a possible threat.

14. How to make complete rounds (premises check).

15. Practical and written exam.

Final Meeting With Executive Staff:

1. Final recommendations on lighting and surveillance cameras.

2. Review procedures for evacuation and situation room.

3. Present an overview of all that has been covered.

4. Present any other recommendations that have been observed and noted.

5. Leave follow-up book for future reference.

AIM OF THIS MANUAL

Everyone in your organization should be trained in what to do in emergency situations regarding every aspect of Anti-terrorist Private Sector Security. As important, they must know what to look for to prevent a terrorist act. Proper training gives them the confidence to do their job completely, thoroughly and with a pride and self-assurance previously lacking. You must have a competent security team and a staff that knows exactly what to do in an emergency situation. You can also rest assured that because of the nature of anti-terrorist private sector security, its reliance on intense alertness and observation, noting people, objects, out of place vehicles, that **criminal activity** will almost be **non-existent**. Your staff and employees will now feel safer in the work place because you have done everything humanly possible to make your business secure from terrorist activity.

For more information on **The Knight's O' the Round Table,** contact the following phone numbers or e-mail addresses:

Phone numbers:

(757) 343-3163
(781) 316-6607
(304) 325-6134

E-mail:

action4two@msn.com
skygodknight@yahoo.com

FAX:

(304) 323-3516

SCOTT KNIGHT

Scott Knight was born in Pennsylvania and now resides in West Virginia with his wife, Rose. Scott served ten years in the U.S. Army, including one year in Mechanized Infantry, and six years in Special Forces. While in Special Forces, he was assigned to HHCIMA at the JFK Center as a Senior Instructor in Phase 1 of S. F. Q. C., teaching all aspects of basic skill courses for future Green Berets (land navigation, survival, patrol techniques, hand-to-hand combat and airborne/air mobile operations. He also served two and a half years in Alpha Company of the Commander-in-Chief Guard. In 1987 he began working in the private sector as a Security Specialist in Strike Security and Executive Protection. He is a certified instructor in Firearms, Monadnock Baton, OC (pepper spray), and hand-to-hand combat. In 2000, Scott, along with Ron Gaunt, formed "Knight's 0' The Round Table" which specializes in teaching Strike Security, Executive Protection and the training and use of K-9 Units. The "Knight's" have handled strikes and provided bodyguard and K-9 services from the coalfields of West Virginia to larger, metro area cities such as New York and Los Angeles. Following the September 11, 2001 disaster, Knight revamped the company's agenda. Recruiting experts in explosives, first-aid, computers, anti-terrorist security, firearms and K-9 they retooled to become a complete Anti-terrorist Security Company. Their primary goal is to help business and industry better prepare against terrorism by training and advising companies in anti-terrorist tactics, the ways and means to make companies a "hard" target and how to react to a terrorist attack if targeted. "The Knight's 0' the Round Table" fill a void in the anti-terrorist security realm that the FBI and Homeland Security cannot fill.

RONALD GAUNT

Ron is a 30-year service retiree of Law Enforcement, pursued after his time in the U. S. Navy where he was assigned to "NAGDAF" from 1962 through 1966. Areas of assignment included Robbery/Homicide, Narcotics/Intelligence, Patrol, SWAT, Special Enforcement and K-9. He has served as instructor in basic baton, advanced baton, basic PR-24, advanced PR-24, OC (pepper spray), crowd and riot control (DART).

He participates in ongoing training in Counter-Intelligence, first responder, Improvised Explosive Devices (IED), explosive detection and recognition, labor negotiations, labor dispute, NLRB defense, and K-9 training and deployment. President of *Travel'n Mandogs Inc.,* he trains and deploys K-9 teams for Fortune 500 companies in patrol, tracking and detection of explosives, narcotics and cadaver (website: travelnmandogs.com). He is a certified Master Trainer, Kennel Master and has worked with service dogs for over 20 years, receiving numerous awards.

Education:
AS degree in Police Science
BS degree in Public Management/Business Management (Pepperdine University, Los Angeles)

Synopsis

The enemy is already at the doorstep. We've been lucky—to date, only one strike has hit our shores (from which we are still recovering). I told the FBI in 1996 that we were vulnerable. I didn't know when, or what would happen, I just knew we would be hit. Truth be known, though, I never imagined the magnitude of that hit. I am not a prophet, nor a seer, but studying terrorism and terrorist tactics since the 1960s and having also studied Moslem Extremism as well as the true Islamic Religion I knew we were on their list. I'm simply an American concerned about the welfare of our citizens, business, industry and way of life.

We have been invaded and it's not close to being over. They are dormant now but are planning, of that you can be sure. They are doing their surveys, analyzing, and picking targets. The only problem is, they will pick an attack more devastating than September 11. There will be smaller attacks (a terrorist must act to justify his very existence) but unless states and the government act, we will at their mercy. At this time computer capabilities, faxes, and along with intrastate controversy, ethnic challenges and the ability to move freely make it easy for the terrorist to move constantly.

States can combat the threat of terrorism, but it will require a rethinking of private security measures, training and mindset of security personnel as well as the reorganization of security assets. To some extent this need has been augmented by the Department of Homeland Security, but it will be years before we feel its effects on the state and local levels. Until then, it is up to us, and we must react and get involved as citizens. The New War is here; it's on our shores and in our midst.

It will not be a war like the first two world wars, Viet Nam, or Desert Storm. It is not a war that will be won with military might, nuclear power or great military strategists. It will only be won through the diligent watch and involvement of our citizenry. I have complete confidence in our military; they can take care of their own. They have developed teams to deal with terrorists—we haven't.

As a nation, and a people, we expect Law Enforcement and the Military to protect us. Welcome to the New War. We have been complacent and lethargic in our lifestyles. We must change this and become, not only involved, but also be aware, of our surroundings, the very common act of going to our jobs. This will be a war in which every citizen must become a soldier, not in the sense of carrying a gun or wearing a uniform but as part of a vast intelligence network concentrating on surroundings and reporting to authorities.

Make no mistake about it, we are on the verge of losing some of our civil liberties because of September 11, sad but true. To save them, we, as a people, as a nation, as a business, as an industry, as a family must react and **BE AWARE**. If we don't, we face the same situation as Israel. We have not had suicide bombers yet, but believe the obvious, they are out there, ready and willing.

Too many people have died for our life, liberty, and happiness to give in to a handful of maniacs. As a last note I would like to leave you a speech made by another patriot. I hope you will read it carefully and look at the insight it presents for our predicament today.

"The unity of government, which represents you as one people, is also dear to you—it is justly so; for it is a main pillar in the edifice of your real independence; the support of your tranquility at home, your peace abroad, of your safety, of your prosperity in every shape of that very liberty, which you so highly prize. But, as it easy to foresee, that, from different causes, and different quarters, much pain will be taken, many artifices employed. To weaken, in your minds, the conviction of this truth—as this is the point in your political fortress against which the batteries of **INTERNAL AND EXTERNAL** enemies will be most constantly and actively. **THOUGH OFTEN COVERTLY AND INSIDIOUSLY** directed, it is of infinite moment, that you estimate the immense value of your National Union to your collective and individual happiness. You should cherish a cordial, habitual and immovable attachment to it. Accustoming you to think and speak of it as the palladium of your political safety and prosperity, watching for its preservation with jealous anxiety.

Discounting whatever may suggest even a suspicion, that, in any event be abandoned and indignantly frowning upon the first dawning of every attempt to alienate any portion of our country from the rest, or to enfeeble the sacred ties which now link the various parts.

For this you have every inducement of sympathy and interest. Citizens by birth or choice of a common country, that country, has a right to concentrate your affections—the name of American, which belongs to you, in your Nationalism, must always exalt the just pride of patriotism, more than any appellation derived from local discriminations. With slight shades of differences, you have the same religion, manners, habits and political principles. You have in a common cause, fought, and triumphed together; the independence and liberty you possess are the work of joint councils and joint efforts—of common dangers, sufferings, and successes.

But these considerations however powerfully they address themselves to your sensibility are greatly outweighed by those, which apply more immediately to your interest. **HERE EVERY PORTION OF OUR COUNTRY FINDS THE MOST COMMANDING MOTIVES FOR CAREFULLY GUARDING AND PRESERVING THE UNION OF THE WHOLE."**

This is a portion of the farewell speech given by the first President of the United States, George Washington, given in September 1789, upon his leaving office. In his foresight he knew we would have enemies and knew that the preservation of democracy would depend on the very citizenry that founded it. As a nation, we now have a common bond and our unity and dedication for preserving the rights and freedom so many have died for will be tested. It is time for us to put aside political divisions, economic divisions and social and class distinctions. We are all in this together. Freedom and democracy is not a free gift—it is a way of life we believe in, a concept our forefathers developed and died for, an idea we live by and die for, and now it is time again to unite and protect the very substance of what we stand for—**ONE NATION, UNDER GOD, INDIVISIBLE, WITH LIBERTY AND JUSTICE FOR ALL**.

Loren Jackson

ABOUT THE AUTHOR

Loren Jackson was born and raised in Belle, W. Va., a suburb of Charleston. After graduating from Dupont High School he went on to earn a Bachelor of Science Degree from Concord University in Political Science. It was here he first started his study of terrorist's and terrorism. In a class on Political Theory he wrote a paper on Terrorism and it's Future in the Western World. After college he maintained an interest in middle-eastern terrorism while serving as a high school history teacher and football coach. He has followed the terrorist groups and the Islamic Religion from the PLO, through the terrorism of Illyich Ramirez Sanchez (Carlos), the Abu Nidal Organization, to the present al-Qaeda and bin-Laden terrorism. His concern is in the protection of this country and its civil rights, as well as the training of our security forces to deal with an enemy within. He has dealt in violent strike security, executive protection and anti-terrorist security.

www.ingramcontent.com/pod-product-compliance
Lightning Source LLC
Chambersburg PA
CBHW022103170526
45157CB00004B/1456